BEYOND THE PILL

BEYOND THE PILL

A 30-Day Program to
Balance Your Hormones,
Reclaim Your Body, and
Reverse the Dangerous Side
Effects of the Birth Control Pill

JOLENE BRIGHTEN, NMD

HarperOne
An Imprint of HarperCollinsPublishers

HarperOne

This book contains advice and information relating to health care. It should be used to supplement rather than replace the advice of your doctor or another trained health professional. If you know or suspect you have a health problem, it is recommended that you seek your physician's advice before embarking on any medical program or treatment. All efforts have been made to ensure the accuracy of the information contained in this book as of the date of publication. This publisher and the author disclaim liability for any medical outcomes that may occur as a result of applying the methods suggested in this book.

Image on p. 28 courtesy Tefi/Shutterstock.
Image on p. 107 courtesy gritsalak karalak/Shutterstock.
Image on p. 201 courtesy Olga Zelenkova/Shutterstock.

BEYOND THE PILL. Copyright © 2019 by Jolene Brighten, ND. All rights reserved. Printed in the United States of America. No part of this book may be used or reproduced in any manner whatsoever without written permission except in the case of brief quotations embodied in critical articles and reviews. For information, address HarperCollins Publishers, 195 Broadway, New York, NY 10007.

HarperCollins books may be purchased for educational, business, or sales promotional use. For information, please email the Special Markets Department at SPsales@harpercollins.com.

FIRST EDITION

Designed by SBI Book Arts, LLC

Library of Congress Cataloging-in-Publication Data has been applied for.

ISBN 978-0-06-284705-8

19 20 21 22 23   LSC   10 9 8 7 6 5 4 3 2 1

To every little girl who's been told she talks too much or asks too many questions.

And to all the women who went before us so that we could see there is a better way.

CONTENTS

PART I

THE PROBLEM WITH THE PILL

CHAPTER 1

REAL TALK ABOUT THE PILL

Angela hadn't had her period in over a year. After months of progesterone withdrawals, blood tests, an MRI, and one recommendation after another to either go back on the pill or move forward with IVF, she made her way to my office as a last-ditch effort in hopes of seeing her cycle return. She had stopped using the pill more than a year before she decided to try to become pregnant. But the end of the pill also meant the end of her periods—and for Angela, this meant no hope for a baby. After more than a decade on the pill, her body no longer followed its original rhythm.

Angela began the pill after her doctor offered it up as a solution for her menstrual pain and suggested that it would help control her acne too. At age sixteen, the pill seemed like a gift—her periods were easier and lighter, and her skin looked better. But at twenty-nine, Angela grew concerned—she knew she could always try for a baby with IVF, but she had to wonder: Just what had happened to her periods? Why had her cycle changed so drastically?

What Angela's doctor didn't tell her is that some women *never* see a return of their periods after stopping the pill. It's known as post-pill amenorrhea and is defined by missing a period for more than three months after stopping the pill if you had regular cycles before the pill, or six months if your cycles were irregular. In Angela's case, her doctors should have seen red flags when she went to them after four months of no period.

While as many as 40 percent of women who come off the pill experience menstrual irregularities like post-pill amenorrhea or short luteal phase (frequent periods), regardless of how long they are on the pill, most women are never told about this risk.

It's not the only risk your doctor isn't telling you about.

If you've picked up this book, you are likely questioning your relationship with the pill. Are you thinking about starting the pill? Or getting off the pill? Either way, I've got your back. Do you want to know how the pill really affects your body (*psst,* it ain't good) or what you can do to minimize your risk while on it? Or are you wondering if some of those alarming side effects they list on the packet (like stroke!) are true and if there is anything you can do to protect yourself? Yes, they are true, and I'm here to tell you that those side effects aren't the only ones. Are you worried that if you stop taking the pill you will have horrible periods or acne again? Or did you already ditch your birth control pill but instead of blissful, predictable periods you are now dealing with hormonal chaos? Yikes! I see this in my clinic every day, and I have to tell you, it's not that uncommon for periods to come back with a vengeance: heavy, painful, and with raging acne or wild mood swings. If any of these scenarios sound familiar, then this book is for you.

Whatever your relationship with the pill, *you are in the right place.* Wherever you are on your journey, *this book will provide you with the best tools to help you.* Straight out of the gate, I want you to know I'm not anti-pill. Nope. I'm pro–informed consent, which means doctors giving you all the information you need to consent to taking birth control.

I spent more than a decade on the pill and know all too well the symptoms of hormonal chaos that follow when you stop the pill for good. **More than half of us are put on the pill for reasons other than preventing pregnancy, and ending that relationship can result in some serious period problems.** This book will help you understand what those hormonal symptoms mean, how the pill is affecting your body, and what to do if you either need to stay on it or are ready to get off it. You're about to discover how to kick your unwanted hormone symptoms, reclaim your hormone health, and make friends with your period—all in 30 days.

In This Chapter

- The down-and-dirty facts about the pill

- Why the pill won't fix your period

- Why post–birth control syndrome (PBCS) might be your worst nightmare

- How the 30-Day Brighten Program can reset your hormones

- What your hormones are telling you

The Pill: The Good, the Bad—and the Ugly

Without a doubt, the pill has benefitted women tremendously. I want to be clear: there's no judgment from me if you're currently on the pill. (As I just mentioned, I took it for a long-ass time myself!) The advent of the pill was revolutionary, and it resulted in sweeping social and economic improvements that gave women the freedom to choose whether or not to have children and work outside the home. It quite literally changed women's lives. In fact, one study found that the pill was partly responsible for an estimated 30 percent increase in women's wages by the 1990s. Access to the pill also contributed to higher college enrollment and completion rates among women in the 1960s and 1970s.

While I strongly believe women should have access to all forms of birth control, I also think every woman should have a resource to turn to in order to decide what is best for *her* body and reduce her risk if she starts taking the pill. The truth is, the pill hasn't kept up with the times, and there's been a bit of an "If it ain't broke, then don't fix it" attitude about it. Pharmaceutical companies have very little incentive to improve upon the pill because they know the burden of pregnancy falls on us. We are women—we cultivate, gestate, and birth life, which is pretty badass but can also hijack our aspirations if we're not ready. Which is why we all can be grateful for a means to prevent pregnancy, but this also leaves us asking, "Why haven't we seen more advancements in this rea?" Women have complained of side effects since the introduction of the ill in the 1960s, but often these concerns are dismissed, and studies as well as doctors have told us that what we're experiencing simply isn't true. (To be fair, the majority of studies have only been able to demonstrate an association of the pill with symptoms, not causation.) There's an unspoken truth in all of this: women will tolerate even the worst side effects in order to maintain their freedom. Some of these side effects are serious—and many of them you probably don't even know about, just like Angela, because no one has brought them to your attention. Until now.

The Risky Business of the Pill

You've probably heard that the pill can make you gain weight or cause you to be "moody." Well, a study in the *Journal of the American Medical Association* of over one million women showed that women who began the pill were more likely to be prescribed an antidepressant—which means it contributes to a bit more than moodiness. Maybe you also read about the risk of stroke on the package, which warns against taking the pill if you're over thirty-five or a smoker. But did you know the pill is also associated with an increased risk of autoimmune disease, heart attack, and thyroid and adrenal disorders?

Unfortunately, too many women are prescribed the pill for non-contraceptive reasons—painful periods, endometriosis, PCOS (polycystic ovary syndrome), acne—without being told of the possible ramifications on their body and hormones. Your doctor really does want you to feel better; it's just that they're taught to see the pill as a simple solution to [fill in the "lady trouble"]. If you've been prescribed the pill for something other than birth control, maybe you heard that inner voice whispering, *This doesn't seem right*. Well, that voice is on to something.

What I see in my clinical practice every day are women with hormone imbalances, looking for a solution—any solution—to their symptoms beyond the conventional dangerous options of birth control, hysterectomies, and IVF. They want a root-cause solution and a better understanding of how to take care of their bodies. This is what I treat, and I'm so glad you've picked up this book, because that solution is now in your hands. This book is for women who've had their hormone symptoms dismissed by their doctors or are looking to take ownership of their health and bodies. Women who've been told, "The side effects are so minimal. Why are you even asking about them?" The pill will not fix hormone imbalance, and you're right to question that. If you're one of these women suffering needlessly with hormonal imbalance, or taking the pill for that imbalance but wondering how you might treat your symptoms more naturally, then this book is for you.

Maybe you started taking the pill solely for birth control, so you're probably fine, right? Nope. This is not a "that other girl" problem. This is a pill problem, and it affects every woman at some point in her life. Which is why I'm going to help you to recognize the side effects, minimize them, and support your body so well that you can get off the pill when you're ready with fewer issues affecting your health and fertility.

If you're thinking about taking the pill or already have been on it, I want you to know about the potential risks to your health. You'll learn more about these as you move through each chapter, but I'll highlight some of them for you now. Symptoms of the pill include:

- hormonal confusion: missing or irregular periods, light or heavy periods, short cycles, infertility, headaches
- digestive problems: leaky gut, gut dysbiosis, inflammatory bowel disease
- energy reduction: fatigue, adrenal and thyroid dysfunction
- skin issues: hair loss, dry skin
- mood disruption: depression, anxiety
- lady part disturbance: low libido (Oh, hell no!), vaginal dryness, chronic infection, pain with sex
- vitamin, mineral, and antioxidant depletion (such as folate, B12, and magnesium)

While these side effects certainly aren't good, they are only the beginning of the potential damage the pill may cause your body. The pill also:

- sabotages your thyroid
- intensifies the risk of blood clots, which lead to strokes
- increases the risk of breast, cervical, and liver cancers
- increases the risk of diabetes

- raises the risk of heart attacks
- triggers autoimmune disease

Are you concerned yet? Because you should be. Extended use of the pill also destroys your gut integrity, causing inflammation, altering your microbiome, and ultimately creating so many issues with immune regulation that it may spark an autoimmune disease. In fact, there is a 300 percent increased risk of developing Crohn's disease if you take the pill. Clearly we need more research into the long-term health consequences of the birth control pill so you'll know what you're getting into when you pop it daily. But don't worry, because this book is going to help you undo that damage, even if remaining on the pill is the only option that's right for you.

Why Are You Really Taking the Pill?

More than half the women prescribed the pill are taking it for reasons other than to prevent pregnancy. The pill is regularly handed out by doctors as *the* solution to any "female" problem. While almost 60 percent of women taking the pill do so for *symptoms,* doctors often don't bother to ask *why* their patients are having these symptoms, or investigate the root cause. The result? A condition that could be treated is left to silently progress, while being masked by the pill, plus these women live with the pill's dangerous side effects. A doctor prescribing the pill to address a hormone symptom is like taking a painkiller for a splinter instead of just removing the splinter. It might make you feel better for a bit, but eventually you're going to feel that splinter again. When a doctor prescribes the pill to a woman without questioning the underlying cause of her symptoms, he or she is doing her a great disservice.

Some of the most common reasons women take the pill include the following:

- menstrual cramps or pain (31 percent)
- irregular, sporadic periods; period problems (28 percent)
- acne, skin conditions (14 percent)
- endometriosis (4 percent)
- unspecified hormonal symptoms (11 percent)

Women also take the pill for headache and migraine relief, abnormal hair growth or loss, PMS (premenstrual syndrome), PMDD (premenstrual dysphoric disorder), mood swings, and mood disorders like depression and anxiety.

I'm here to tell you that the pill won't fix your period. It may seem like a magic pill at first when those painful periods are more manageable and you're not going through a supersize tampon every hour. But at what risk to your health in the long run? Even worse, the pill can create a dependency if you believe your only option for avoiding hormone symptoms is to continue taking it. The trouble is the pill masks symptoms instead of addressing the root cause of them. Symptoms are your body's way of trying to tell you something. And you need a root-cause resolution, not another hormonal Band-Aid! In this book, I will teach you how to identify what these symptoms are telling you so you can correct your underlying hormone imbalance. You don't have to be stuck on the pill, or living in fear of what might happen when you come off it.

What Is Post–Birth Control Syndrome?

The pill essentially shuts down the conversation between your ovaries and your brain, so it's no surprise that once you stop taking it you may encounter challenges with reestablishing that connection. It's kind of like when you block someone on your phone—communication is not an option. Well, when you take the pill, it's as if the brain is blocking the ovaries, but only because the pill told it to. If you've been taking the pill to treat symptoms like PMS, heavy periods, acne, or mood swings, I hate to break it to you but there's a good chance you'll be dealing with those symptoms again. That is, unless you do something about them.

Post–birth control syndrome (PBCS) is a constellation of symptoms women experience when they discontinue hormonal birth control. The women I see in my clinic suffering from PBCS often experience the return of the very hormone symptoms that drove them to take the pill in the first place—plus some added ones the pill has caused that they just weren't aware of yet, such as a loss of a period altogether or first-time adult acne (fun!). PBCS symptoms can range from hormonal irregularities, which may include loss of menstruation, infertility, pill-induced PCOS, and hypothyroidism, to gut dysfunction and autoimmune symptoms. PBCS generally occurs within the first four to six months after discontinuing the pill and, in my experience, doesn't just go away without taking the necessary steps I will teach you in this book.

My Pill Story

I see patients like Angela every day. Unfortunately, many patients come to me after they've been told that there's nothing they can do about their condition—or worse, that there's nothing medically wrong with them and they should just go back on the pill. But your symptoms should not be ignored. You are the only one who knows what "normal" is for you.

I have my own pill story. We all do, right? Either the pill left you with constant rage or you gained 10 pounds you just couldn't lose. Why have we all accepted this as the normal consequence of preventing unwanted pregnancy or healing symptoms as simple as acne? Well, I understand the emotional and physical dependencies, worries, and exhaustions you're going through: I spent ten years on the pill and endured a lot of grief as a result. I really wish someone would have told me just how badly one little pill could wreck my body.

I first fell in love with the pill when my seven-plus-day vomit-inducing periods went away. That was until I started crying all the time, experienced vaginal dryness, endured relentless yeast infections, and developed pain with intercourse (vaginismus, if you want to get technical). I was only twenty-one, and I didn't know why my body was constantly betraying me. My doctors never told me it could be the pill. And get this: rather than talking to me about the pill, my doctors recommended I have surgery to cut the nerves in my pelvis so I wouldn't have pelvic pain anymore. WTF! Seriously, WTF. I even became allergic to Monistat due to overuse for chronic yeast infections. No joke: my vagina basically turned inside out and I wanted to die. And not once but twice I found myself in the ER thinking I would die due to a ruptured ovarian cyst. I wish my doctors would have said what is so obvious to me now: the birth control pill that was sold as something to ease my symptoms was messing with the hormone balance in my body and making my symptoms worse. But instead my doctors recommended that I cut off any feeling to my pelvis and fill that Diflucan prescription, and reassured me that a libido wasn't necessary.

When I finally ditched the pill, my period vanished, I suddenly had adult acne, and I was filled with rage, blowing up at loved ones, storming out of rooms, and hating on anything that looked at or talked to me. I was an emotional wreck. It was so bad I basically hid from society in order not to destroy every bit of the social network I had. My doctor was unconcerned by my missing period, which before the pill had been regular, and he was even less concerned that I might someday want to have a baby. His only solution? Get back on the pill. When my period finally did come back, I found myself filled with adolescent PTSD as I bled through my pants and the cloth chair I sat on in class. You'd think I might be slightly less embarrassed sitting in a room full of future doctors, but nope. It sucked just as bad as it did when I was fourteen. These are all classic symptoms of PBCS. And with the very diet, lifestyle, and supplement protocols I'll give you in this book, I was able to restore my period, clear my skin, revive my libido, and become a mother to one beautiful boy.

But being on the pill wasn't all bad. I'm a first-generation college student, and I used to joke that all men had to do was *look* at a woman in my family and she'd get pregnant. The pill was instrumental in my academic and professional achievements. I was able to make the choice to have a baby when I was ready. I cannot understate the freedom we women experience when we can control our fertility. And while the pill did nothing to address the root cause of my symptoms, it sure was nice to get a break from a period that kept me from going to school and socializing in general.

Through my journey of healing my body, balancing my hormones, and getting my period back, I came to understand what it means to struggle with health concerns . . . and what it feels like to have your symptoms dismissed or ignored. Now I'm considered one of the leading experts in post–birth control syndrome and have helped thousands of women who struggle with hormone imbalance. I want to help you too. I'm one of the first clinicians to spend countless hours researching PBCS to come up with appropriate diet and lifestyle interventions. I have a background in nutritional biochemistry—nerd alert!—and love research, data, and journal articles (I read them every weekend). Before earning my doctorate in naturopathic medicine, I earned degrees in chemistry, nutritional biochemistry, and clinical nutrition, so my work is grounded in both scientific research and clinical experience.

There is a root cause to your symptoms, and I want to help you dig deep to find it. In my clinic, my motto is "I don't heal my patients—I teach them how to heal themselves." I will help you do the same. Working with patients in my medical clinic, I developed the 30-Day Brighten Program, which has helped thousands of women struggling with the side effects of the pill, including PBCS.

The 30-Day Brighten Program

Beyond my experience as a woman who broke up with the pill, I want to share with you what I've learned while creating this effective and comprehensive program that helps women undo the effects of PBCS and eliminate their dependency on the pill. (Yes, you don't need to suppress your hormones every day to feel amazing in your body.) The 30-Day Brighten Program you'll find in chapter 12 is designed to reverse the harmful effects of the pill with diet and lifestyle interventions, and a targeted supplement plan to individualize your experience. You'll learn the essentials of how to come off hormonal contraceptives, restore your hormonal health, and get your badass self back! **The program works not only for women who elect to switch to a non-hormonal form of birth control but also for those who opt to remain on the pill yet want to reduce the risk factors and side effects.**

Before you begin the program, in part I of the book you'll explore PBCS and its many symptoms. I'll help you understand how your menstrual cycle works and the role of various hormones in your body as well as how the pill disrupts this cycle. We'll look at what your period may be trying to tell you about your hormonal health, and I'll give you insight into which lab tests to have for period problems and other hormonal symptoms. Because I want you to be able to start feeling better as soon as possible, while you wait for those lab tests, I will offer up some natural solutions you can implement right now.

In part II, you'll take a deeper look at the key systems the pill disrupts and how each can contribute to the symptoms women experience when on the pill and during PBCS. You'll discover how the pill affects your liver, gut, thyroid, adrenals, metabolic system, mood, libido, and fertility, and you'll learn how to support your body whether you continue with the pill or choose to give it up altogether. That's why I've included easy-to-implement solutions within each chapter that will both lower your risk if you choose to remain on the pill and resolve your symptoms if you're coming off it. What you'll find in part II are the advanced protocols, or, in other words, some next-level healing. And if you're taking the pill for any reason besides pregnancy prevention, these chapters will guide you toward understanding your symptoms and show you how to break free from your dependency on the pill.

In part III, you'll learn about my concrete strategy for protecting your body on the pill and when coming off the pill, including alternative forms of pregnancy prevention. You'll also begin the 30-Day Brighten Program, which I've been using in my clinic to support women in reversing PBCS and taking back their bodies. This program was developed with feedback from my patients, and with it you will become your own health detective, determining what your body's signals mean and how to best address them naturally with a whole-foods, anti-inflammatory diet, designed to resolve nutrient depletions and support hormone balance. Plus, you'll create a personalized supplement plan using my guide, and you'll ensure your success in the program using the meal plans and delicious recipes I've included. Because who needs the stress and frustration of trying to figure out what to eat every day?

Here's more of what you'll learn:

- What you need to know about being on the pill and how to support your body

- How to eliminate symptoms of PBCS (hormonal headaches, acne, painful periods, hair loss, depression, anxiety, and more)

- What every woman must know about getting pregnant after the pill

- How to promote a better mood, greater energy, and easy, predictable periods

- The major systems in the body the pill and other hormonal contraceptives affect

- Supportive therapies to naturally detoxify your liver

- What to eat to balance your hormones, heal your gut, and reverse an autoimmune disease

- Supplements that every woman should consider in coming off the pill and what you must take if you stay on it

- Tools to reset your circadian rhythm, reduce stress, and sleep soundly

- How to say buh-bye to PMS and excess weight, and hello to libido

- Fertility-boosting strategies and tips to optimize conception

- Healthy ways to reduce your risk of cancer, stroke, and heart attack
- All about the fertility awareness method (including stats on its accuracy)

This book is aimed at educating, empowering, and supporting you to make the best decisions for your health. **Ultimately, *you* are the only person who can decide the best form of birth control for you—after all, you are the only person living in your body.**

As for my patient Angela, her period returned (without any annoying symptoms like PMS, rage, or cramps) after she followed the very program I'm going to share with you. The program's methods have helped thousands of women regulate their cycle, love their period, and be healthy and happy. And Angela got pregnant! By restoring her body's health and undoing the effects the pill had had on her body, she was finally able to have the baby she had been longing for.

How to Use This Book

I'm a firm believer that if you're going to present a gal with a problem, you need to follow that with a solution. Real talk—I'm going to present you with a lot of problems caused by the pill, some of which are legit scary. But I'm also going to provide you with solutions, so you can take action immediately. If you've picked up this book, then you need the 30-day program to get you to that state of hormone bliss that deep down you know is possible. This program works for women on the pill, transitioning off the pill, struggling with PBCS, or wanting to avoid using the pill to "fix" their period problems. This is your starting place and the number one thing you need to do before anything else.

Throughout these chapters you'll find protocols to address the issues you specifically may be experiencing. Not every woman has the same side effects from the pill, and not every woman has the same symptoms in PBCS. With that in mind, I've provided you with solutions throughout part II that you can build into the 30-Day Brighten Program or you can use to fine-tune after the thirty days. I provide guidance on how to customize the program to fit your needs, and it is my sincere hope that you'll revisit the protocols and the 30-day program whenever symptoms arise. My goal is to empower you and put the medicine in your hands with the protocols.

Those protocols are:

The Brighten Detox Protocol

The Brighten Gut Repair Protocol

The Brighten Thyroid and Adrenal Health Protocol

The Brighten Metabolic Protocol

The Brighten Mood Mastery Protocol

The Brighten Libido-Boosting Protocol

The Brighten Supplement Protocol

Your Hormone Health Journey

Are you ready to get started on your transformation to create ah-mazing hormone health? Take the following Hormone Quiz to discover which hormonal imbalances you might be struggling with right now. When you reach part III and are ready to begin the 30-Day Brighten Program, return to this quiz so you can customize your lifestyle and supplement plan based on your needs. I recommend taking the quiz again after completing the program; this will help you track your symptom changes and troubleshoot the next steps in your healing journey.

After you finish the quiz, the next chapter will give you the lowdown on your hormones. Once you understand them and how they can help you, it's even easier to start working with your body to reset them.

Hormone Quiz

As you read through this quiz, check the symptoms that apply to you currently, then total up your symptoms in each category.

Category A

- ☐ I experience bloating or puffiness.
- ☐ I feel irritable or experience mood swings.
- ☐ I experience heavy, painful periods.
- ☐ I have gained weight or have difficulty losing weight, especially around my hips, butt, and thighs.
- ☐ I've been told I have fibroids.
- ☐ I sometimes cry over nothing.
- ☐ I get migraines or other headaches.
- ☐ I have brain fog.
- ☐ I've had gallbladder problems or had my gallbladder removed.

TOTAL _____

Category B

- ☐ I'm emotionally fragile and/or I feel nostalgic about the past.
- ☐ I have difficulty with memory.
- ☐ My periods are fewer than three days.
- ☐ I struggle with depression, anxiety, or lethargy.
- ☐ I have night sweats and/or hot flashes.
- ☐ I've had trouble with recurrent bladder infections.
- ☐ I sometimes have problems with urinary leakage.
- ☐ I have difficulty sleeping and wake at night.
- ☐ My breasts are smaller and/or beginning to droop.
- ☐ I have achy joints or am prone to joint injuries.
- ☐ My sun-damaged skin is more noticeable.
- ☐ I am noticing more fine lines and wrinkles.

Hormone Quiz

- ☐ I have dry or thinning skin.
- ☐ I have no interest in sex.
- ☐ I have vaginal dryness or pain with intercourse.

TOTAL _____

Category C

- ☐ I experience PMS seven to ten days before my period.
- ☐ I get headaches or migraines around my period.
- ☐ I feel anxious often.
- ☐ I have painful, heavy, or difficult periods.
- ☐ My breasts are painful or swollen before my period.
- ☐ I feel agitated, irritable, or weepy before my period.
- ☐ I have had a miscarriage in the first trimester.
- ☐ I experience restless legs, especially at night.
- ☐ I have had difficulty getting pregnant (after trying for six or more months).

TOTAL _____

Category D

- ☐ I have abnormal hair growth on my face, chest, and/or abdomen.
- ☐ I have acne.
- ☐ I have oily skin and/or hair.
- ☐ I have areas of darker skin (e.g., armpits)
- ☐ I've noticed thinning hair on my head.
- ☐ I have skin tags.
- ☐ I struggle with depression and/or anxiety.
- ☐ I have PCOS.
- ☐ I have had difficulty getting pregnant (after trying for six or more months).

TOTAL _____

Hormone Quiz

Category E

- ☐ I have a low libido or diminished sex drive.
- ☐ I struggle with depression, have mood swings, or cry easily.
- ☐ I have no motivation.
- ☐ I am tired or fatigued throughout the day or have been diagnosed with chronic fatigue syndrome.
- ☐ I'm unable to gain muscle, and I'm losing muscle mass.
- ☐ I have a decrease in bone density or have been diagnosed with osteopenia or osteoporosis.
- ☐ I have urinary incontinence.
- ☐ I have a loss of sexual fantasies.
- ☐ I have difficulty or am unable to orgasm.
- ☐ I have cardiovascular symptoms or heart disease.
- ☐ I've had weight gain.
- ☐ I have anxiety or panic attacks.

TOTAL _____

Category F

- ☐ I feel tired in the morning, even after a full night's sleep.
- ☐ I depend on caffeine to get through my day.
- ☐ I want to take naps most days.
- ☐ My energy crashes in the afternoon.
- ☐ I crave salty or sweet food.
- ☐ I'm dizzy when I stand up too quickly.
- ☐ I feel at the mercy of stress.
- ☐ I have difficulty falling asleep and/or staying asleep.
- ☐ My muscles feel weaker.

Hormone Quiz

- ☐ I get sick often and/or have a difficult time getting over infections.
- ☐ I have low blood sugar issues.

TOTAL _____

Category G

- ☐ My life is crazy stressful.
- ☐ I feel overwhelmed by stress.
- ☐ I have extra weight around my midsection.
- ☐ I have difficulty falling or staying asleep.
- ☐ My body is tired at night, but my mind is going a mile a minute— "wired and tired."
- ☐ I get a second wind at night that keeps me from falling asleep.
- ☐ I wake between 2 and 4 a.m. and can't go back to sleep.
- ☐ I feel easily distracted, especially when under stress.
- ☐ I get angry quickly or just feel on edge.
- ☐ I have high blood pressure or a fast heart rate.
- ☐ I have elevated blood sugar or diabetes.
- ☐ I get shaky if I don't eat often.
- ☐ I'm prone to injury and have difficulty healing.

TOTAL _____

Category H

- ☐ I have brain fog or feel like my memory isn't quite what it used to be.
- ☐ I'm losing hair (scalp, body, outer third of the eyebrows).
- ☐ My hair is dry and tangles easily.
- ☐ I'm constipated often and need a stimulant (like caffeine) to get a bowel movement.

Hormone Quiz

- ☐ I'm cold and/or have cold hands and feet.
- ☐ My periods are sporadic or occur more than thirty-five days apart.
- ☐ I have joint or muscle pain.
- ☐ I have dry skin.
- ☐ I have had difficulty getting pregnant (after trying for six or more months) or have had a first trimester miscarriage.
- ☐ I am in a low mood or struggle with depression.
- ☐ I'm tired no matter how much I sleep.
- ☐ I find it difficult to break a sweat.
- ☐ I have recurrent headaches.
- ☐ I have high cholesterol.
- ☐ I have a hoarse voice most days.

TOTAL _____

Answer Key

0 or 1 box checked in a category = This category is unlikely to be a culprit in the symptoms you're experiencing.

2 to 4 boxes checked = This area needs your attention.

5 or more boxes checked in a category = This just might be your troublemaker, meaning right now this is likely the dominant hormone aggravating your symptoms.

Category A: Too Much Estrogen
Be sure to read chapter 4 to find out more about estrogen dominance as well as page 237 in chapter 12.

Category B: Too Little Estrogen
See page 238 in chapter 12 for more information.

Hormone Quiz

Category C: Too Little Progesterone
  In chapter 4, learn how this relates to too much estrogen, plus see page 240 in chapter 12.

Category D: Too Much Testosterone
  Check out chapter 8 for why you may have this even if you don't have PCOS, and review page 240 in chapter 12.

Category E: Too Little Testosterone
  Learn more about this and its effect on your libido in chapter 10, plus see page 242 in chapter 12.

Category F: Too Little Cortisol
  See page 243 in chapter 12 for what you can do about this.

Category G: Too Much Cortisol
  Check out chapter 7 to see how this relates to your adrenal glands, and visit page 244 in chapter 12.

Category H: Too Little Thyroid Hormone
  You'll want to carefully read chapter 7 if you have any thyroid issues, plus look at page 245 in chapter 12.

CHAPTER 2

THE LOWDOWN ON YOUR HORMONES

Maybe you've heard the story that being a woman is inherently awful—that we have wild mood swings, and our hormones and our bodies constantly betray us. This is a myth and a lie meant to keep us from demanding better for ourselves.

Here is the truth: your body is in *no way* betraying you. In fact, your hormones hold the key to an incredible life. If you're like me, you didn't get the spiel on how your menstrual cycle works, what's going on, and how you can make friends with your period, because sex ed didn't give you any insight into your body or show you a road map to managing symptoms. You may have been taught that your period is even shameful (me too). Then when you see your doctor because you're concerned, he or she may tell you either "There's nothing wrong and it's in your head" or "Yeah, you've got problems and there's only one solution: suppress the hell out of your hormones."

When women stop taking the pill and begin to experience the symptoms of post–birth control syndrome—acne, irregular periods, wild mood swings, or super heavy, painful periods—their doctors often tell them their only option is to go back on the pill. Having always been told that their body is betraying them, they believe that narrative. I'm here to tell you that this narrative is *wrong*. Look, your body has something important to say, but when you're on the pill, it's like your body has been bound and gagged and thrown in the trunk of a car. By the time you pop that trunk and remove the gag, your body is screaming for help because it doesn't want to get stuck in that trunk again. That's essentially what your hormones do when you use the pill just to mask symptoms. But you *can* work with your body and understand your hormonal health. These symptoms are common, but they are not normal.

Your hormones are really important. In this chapter, I'll help you understand how your menstrual cycle works and we'll talk about how your hormones are connected. As much as "that time of the month" can be a real pain in the uterus, your period can give you remarkable insight into your health—if you pay attention. A rhythm occurs at different points in your cycle if you allow your hormones to flow naturally. As part of that, your mood, your ability to get shit done, and your orgasms can vary too. There's a lot more going on in your cycle than just your period.

The pill, on the other hand, causes your hormones to flatline (and that flatline means your orgasms won't be the same!). The pill floods your system with constant doses of synthetic hormones while suppressing others, throwing your entire system out of balance and causing many of the problems we'll explore in this book: blood sugar imbalance, night sweats, headaches, depression, and much more. If you experience any of these symptoms regularly, you're probably in need of a hormone reset.

Is My Period Normal?

If you're like me, your high school sex ed class was an epic fail. Not only did I leave my class without knowing a darn thing about my period or my body, but also I had the bejesus scared out of me about how tragic womanhood is. Luckily, I've devoted my adult life to studying what I should have learned then, so bear with me while I teach you the basics and just how awesome being a woman is.

You've got this awesome rhythm of hormones rising and falling throughout your cycle. When they're balanced and your cycle is flowing smoothly, you shouldn't encounter any major period problems. Menarche—have you heard of it? This is when you first started your period, and it's when you get your va-va-voom estrogen boom! You start to grow breasts and pubic hair. For most women, this occurs between the ages of eleven and fourteen. It's not uncommon to start a little later, but if you haven't had a period by sixteen years old, that warrants investigating. On the other hand, if you got it really early, or what

In This Chapter

- How your menstrual cycle works

- What your period says about your health

- What your teacher missed in sex ed

- How the pill works

- Which tests your doctor should be ordering to evaluate hormone imbalances

we call "precocious puberty," around age eight, that's a sign that you had already been experiencing some hormone troubles.

The typical menstrual cycle lasts twenty-eight days in 10 to 15 percent of women, and can range anywhere from twenty-six to thirty-six days. Cycles are highly individualized, and there is a gradient of "normal" when it comes to menstrual cycles. If your period isn't occurring every twenty-eight days but it's regular, that's okay. There's a spectrum here. It's when you're far outside that spectrum—such as never knowing when your period is coming—that you want to start investigating why. Thank gawd for apps, because before apps most women struggled to track their periods. Now more women are tracking their periods and can more easily identify irregularities.

Day 1 of your cycle is the first day of your period or the first day you see blood. That happens because **estrogen** and **progesterone** drop (for more detailed information on all your hormones, see pages 33–36), which triggers the lining of your uterus, the **endometrium**, to shed, resulting in your period from about days 1 through 7. The drop in these hormones also sets off a cascade: your brain (the pituitary gland, specifically) releases **follicle stimulating hormone (FSH)**, which stimulates the growth of the follicles in your **ovaries** to get an egg ready for **ovulation**. Estrogen begins to rise around day 8, causing your lady parts to plump up, giving you more pronounced curves, and making your lips fuller.

If you notice around day 9 or 10 that you're totally in the mood and you can't keep your hands off your partner, well, that's because **testosterone** is rising at

Menstrual Cycle Glossary

corpus luteum: A mass of tissue formed in the ovary after ovulation that secretes progesterone and eventually disintegrates if the egg is not fertilized

endometrium: The mucous membrane that lines the uterus and thickens in preparation for the implantation of an embryo and is shed during menstruation

estrogen: A female sex hormone produced by the ovaries, dominant in the first half of the menstrual cycle

fallopian tube: One of the two tubes on either side of the uterus that carry the egg from an ovary to the uterus

follicle-stimulating hormone (FSH): A hormone produced by the pituitary gland that stimulates the maturation of the ovarian follicles in preparation for ovulation

follicular phase: The first half of the menstrual cycle when estrogen is high and the ovarian follicles mature in preparation for ovulation

luteal phase: The second half of the menstrual cycle that occurs after ovulation when progesterone is higher

luteinizing hormone (LH): A hormone produced by the pituitary gland that triggers ovulation and the development of the corpus luteum.

ovaries: Female reproductive organs that produce eggs and hormones

ovulation: When the ovary releases an egg

ovulatory phase: The phase between the follicular and luteal phases when LH surges, triggering the release of an egg.

progesterone: A hormone that is released by the corpus luteum in the ovary to help prepare the uterus for the implantation of an embryo, dominant in the second half of the menstrual cycle

testosterone: A sex hormone secreted by the ovaries and adrenal glands that rises before ovulation; highest in men, but necessary in women

Quiz: Menstrual Cycle Myth Buster*

Take the period pop quiz and see how you score. Answer true or false after each statement.

1. Every woman has a 28-day menstrual cycle.
2. A woman can get pregnant any day of the month.
3. A woman can get pregnant only one day of the month.
4. A woman ovulates only one egg per cycle.
5. Women ovulate every single month.
6. Sperm live only one day.
7. The pill regulates a woman's period.
8. Having a period isn't necessary.
9. Day 1 of a woman's cycle is the first day of her period.
10. Progesterone is low in the second half of a woman's cycle.

*See page 331 for the answers.

that time. (Yes, we gals also produce testosterone, just in smaller quantities than our male counterparts.) Your body is super smart. It increases your testosterone and elevates your libido about five days before you ovulate, so that you'll seek out your partner and have sex and retain that sperm—which can live about three to five days—in hopes that once the egg is released, you'll become pregnant. That's why, despite the fact that your egg lives only about twenty-four hours, you're considered fertile for five to six days out of each month. Doesn't it seem kind of silly to suppress your hormones endlessly when you're fertile only about six days a month? Did your doctor ever tell you that? (My doctor sure didn't!) So many women come to my office saying that if they had known how difficult it was to get pregnant, *they never would have started the pill in the first place.* Often it's not until you try to get pregnant that you realize there's this finite window, and getting pregnant or not getting pregnant is more about understanding your specific rhythms than suppressing hormones.

Estrogen is the main player in the first half of your cycle. It spikes around

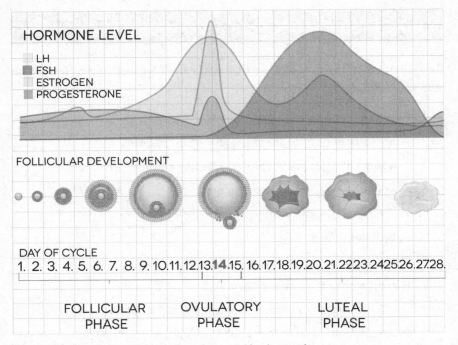

HORMONE LEVEL

- LH
- FSH
- ESTROGEN
- PROGESTERONE

FOLLICULAR DEVELOPMENT

DAY OF CYCLE
1. 2. 3. 4. 5. 6. 7. 8. 9. 10.11.12.13.**14**.15. 16.17.18.19.20.21.22.23.24.25.26.27.28.

FOLLICULAR PHASE   OVULATORY PHASE   LUTEAL PHASE

The menstrual cycle of a woman not using hormonal birth control.

days 12 through 14 to trigger the release of **luteinizing hormone (LH)**, which marks the beginning of the **ovulatory phase** and stimulates your ovaries to release an egg, aka ovulation. It's during this time that the egg travels down the **fallopian tube** and either implants in the endometrium if fertilized by sperm or slowly dissolves and passes out of the body, along with the uterine lining, during your period. Estrogen also stimulates the growth of uterine tissue, thickening your uterine lining for that potential implantation of an embryo. This is why too much estrogen causes heavy, painful periods. It's the predominant hormone during the first half of the cycle, which is known as the **follicular phase** (approximately days 1 through 14 of your cycle).

The ruptured follicle, now called the **corpus luteum**, releases progesterone and estrogen to prepare your body for pregnancy. Progesterone is the hormone that helps you feel chilled out and calm and in love with your life. When that is too low, women will want to either (a) run away into the woods and never be seen again; (b) murder anyone who gets in their way; or (c) do both. If you find that during the week or two before your period you can't sleep, you're feeling anxious, or you're aggravated with everyone in your life, you need to get your

Moon Cycles and Menstrual Cycles

Most women actually follow the moon cycle and get their periods during either the full moon or the new moon. It's believed that you're most fertile when you have your period during a new moon, because ancestrally speaking we would have been up at night two weeks later during the full moon with nothing more to do than get busy with our mate, so it was a perfect opportunity for making a baby. This is known as the White Moon Cycle. The Red Moon Cycle is when your menstrual cycle begins during a full moon and therefore ovulation takes place during the new moon phase. The Red Moon Cycle is thought to be followed by the women who are leaders and healers. Pretty damn cool, right?

See chapter 11 for more about cycling with the moon.

---

progesterone back, and luckily it isn't difficult to do when you take a natural approach like the one in the 30-Day Brighten Program. Progesterone is the most prominent hormone from days 15 through 28 of your cycle and peaks around day 21, which is known as the **luteal phase**.

If the egg doesn't meet sperm, then your hormones drop and your menstrual cycle starts all over again. But if you do become pregnant, those hormones stay elevated.

What Does a Period on the Pill Look Like? (Spoiler: Major Hormonal Confusion)

From progesterone to cortisol to thyroid hormone—there isn't a hormone the pill doesn't disrupt. Now, I'd like to be clear: this chapter (and this book) in no way is meant to condemn you for taking the pill. Instead, my aim is to help you understand how the pill works and how it can impact your other hormones. Knowledge truly is power, and it's time we all had the necessary knowledge about the little pill we take every day. Many women have no idea (through no fault of their own) just how much the pill affects their hormones.

Your hormonal system (also known as the endocrine system) is like a symphony, which under optimal conditions—when every instrument is in tune

Your Cycle at a Glance

- Your menstrual cycle is the number of days from the first day of bleeding in one month to the first day of bleeding in the next month.

- The average length of a menstrual cycle is twenty-eight days, with most cycles falling between twenty-six and thirty-six days.

- The menstrual cycle has three phases: follicular, ovulatory, and luteal.

- A spike in estrogen during the follicular phase causes a spike in LH, which triggers ovulation, typically in the middle of the cycle.

- Following ovulation, progesterone rises and peaks around day 21.

- A fall in progesterone triggers your period if you have not become pregnant.

- The average period is about four to six days in length.

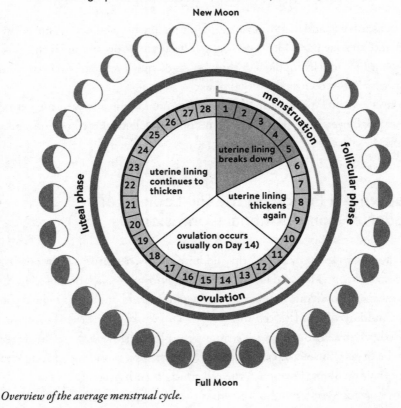

Overview of the average menstrual cycle.

and hitting the right notes—is playing the most beautiful music you have ever heard. But if just one instrument is out of tune, it can ruin the entire piece, making you cringe. The other musicians may begin to change how they're playing in order to compensate for that one out-of-tune instrument, and the music quality begins to diminish further. In this same way, **if one hormone is out of balance, it affects the whole system.**

People, including doctors, are always saying that the pill essentially tricks the body into thinking it's pregnant. Honestly, the body isn't that stupid. When we talk about pregnancy, we're talking about many complex changes. Multiple systems evolve, and you bathe in a wonderful, protective soup of hormones as you grow an entire human. Those pregnancy hormones are nothing like the synthetic hormones offered by the pill.

Remember how your cycle works, with higher estrogen in the first half of the cycle and higher progesterone in the second half? Well, most birth control pills deliver a large dose of both synthetic estrogen and progesterone (progestin, to be exact) throughout the entire month, sans the sugar pill week. This is nothing like a natural menstrual cycle in which hormones fluctuate. This heightened daily dose of hormones suppresses the pituitary from releasing follicle-stimulating hormone (FSH) and luteinizing hormone (LH), which is how it prevents ovulation. The brain perceives there are more than enough hormones, so it diminishes the signal that would demand the body to make more (this is known as negative feedback). That means no brain hormones, no natural hormone rhythm, and no ovulation, which is great if you don't want to have a baby. When you take that week of placebo pills, you're not actually getting a period because you never ovulated. Instead, this is what is called a withdrawal bleed.

There are two types of pills: combination and progestin only. The combination pill contains estrogen and progestin, is more effective, and has less breakthrough bleeding. The combo pill suppresses ovulation, thickens cervical mucus to block sperm, changes tubal motility, and thins the uterine lining. Most women take this type of pill. The progestin-only pill, aka the mini-pill, is typically used by women who have adverse reactions to synthetic estrogen or who are currently breastfeeding. It also stops ovulation but only in about 60 percent of women.

The basic mechanism of how the pill prevents pregnancy is a huge player in the hormonal imbalances it creates. The pill's job is to stop your brain from communicating with your ovaries. As long as you're on the pill, your brain and ovaries aren't talking, which can seriously affect all your hormones.

The Pill and Your "Bad" Boyfriend

Okay, I'm not saying the pill is the sole cause of why you keep dating the wrong guys, but there's evidence that it does you no favors when it comes to mate selection. This is going to take some science and serious nerd talk to explain, but if you stick with me, I promise you'll be fascinated.

You may have heard that your mate's scent is the cause of your attraction. The reason for this is that scent reveals a man's major histocompatibility complex (MHC), a set of genes that play a large role in immune system surveillance. The job of MHC is to take proteins that need to be dealt with by the immune system and display them in an easy-to-find place: the cell surface. Because of its intimate role with your immune cells, your MHC also determines your susceptibility to autoimmune disease.

We know that women preferentially select men who have MHC genes different from their own. Why would we do this? Because a mate with different MHC genes will provide our children with a more varied MHC profile and, therefore, a more robust and well-regulated immune system. It's thought that this evolved as a means to prevent us from being attracted to our relatives, who would offer our offspring less of an advantage in the gene pool.

Ready for the fascinating part? Research has shown that once a woman starts taking the pill, her scent preference shifts toward men who are more *similar* to her genetically, with less interest in those men with differing MHC genes. Crazy, right?

Why might the pill make you more attracted to your cousin? No one really knows for sure. One hypothesis is that since the pill disrupts the hypothalamic-pituitary-gonadal (HPG) axis in a way somewhat similar to pregnancy, this mechanism would make you seek out relatives who might provide you with support.

If the pill causes women to select mates who create babies with less than optimal immune systems, and we've been on the pill for multiple generations, is it any wonder there has been a rise in the incidence of autoimmune disease? Yes, autoimmune diseases are complex in nature and many variables are at play, but a whole lot of us have been on this pill, and we've seen a big rise in autoimmune disease in the generations that followed its induction.

Hormones—A Helluva Lot More Than Just Your Period

Now I'm going to take a deep dive into the major hormones you should know about. There are way more hormones than what are listed here, but that would be an entire book in itself. I want to help you conduct the most beautiful symphony your body can play. Let's go beyond periods and baby making here, because there is a lot your hormones do besides procreation.

To begin with, there are two main groups of hormones: steroid and nonsteroid hormones.

Steroid Hormones

This group of hormones is made from cholesterol, specifically LDL, or "the bad cholesterol." You know, that stuff you're told will kill you if you eat it? Turns out, you can't make your hormones without it, and when you get to reading everything these hormones do for you, you'll quickly see why not eating cholesterol might kill you faster and why a fat-free diet is not your friend.

PREGNENOLONE

I call her the mama hormone because she gives birth to all your steroid hormones. Or rather, all the hormones in this section are made from her. Pregnenolone also depends on other hormones, like thyroid hormone, for synthesis.

ESTROGEN

You may recall from a bit earlier in the chapter that estrogen is responsible for building up the uterine lining so that when the egg is fertilized it has a nice cushy space to call home. Babies aside, this hormone is also how you got all those curves—and how those curves stay perky. Plus, you can thank it for keeping your lips full and your skin looking plump.

But it gets better: natural estrogens help keep your brain firing and your heart in tip-top shape, and are a big reason why your bones don't bend. Estrogen moves throughout your body, affecting nearly every tissue, including your brain, bones, heart, and skin.

Make no mistake: When I talk about how amazing estrogen is, I am *not* talking about that synthetic shit they package up in a pill. I'm talking about what your body naturally makes from your ovaries, adrenals, fat cells, brain, and other tissue. Your doctor may have told you that your pill contains pretty much the same hormones you make naturally. It doesn't.

PROGESTERONE

Progesterone is another key player in fertility, with a big role in maintaining a healthy pregnancy. But even if you don't want a baby, you *do* want to be ovulating. Your ovaries produce progesterone from the corpus luteum, which is only formed after ovulation.

Ever feel anxious, sleepless, irritable, or weepy before your period? That's usually a sign that you aren't getting adequate progesterone. When progesterone is just right, it keeps you feeling chilled out, calm, and in love with your life. Good stuff, right? Progesterone counters the effects of estrogen and has a calming effect, reducing anxiety and increasing sleepiness. It's produced mainly in the ovaries and the adrenal glands (and the placenta when you're pregnant).

Progesterone also helps you use fat for energy, builds and maintains bones, protects your breasts and uterus against cancer, and promotes cortisone production, appetite, and fat storage. And it helps your cells use thyroid hormone.

TESTOSTERONE

Testosterone always gets touted as the male hormone (androgen), but women need testosterone too. Without it, your bones get weak, your brain slows down, your mood tanks, and your motivation goes out the window. This hormone also boosts confidence and your energy. Feeling fatigued? Your doctor needs to check your testosterone.

Testosterone gets us into trouble, like any hormone, when there is too much of it. That's when we can experience symptoms such as oily skin, acne, hair on our face, chest, or abdomen, or loss of hair on our head.

CORTISOL

Cortisol, produced in the adrenal glands, is known as a stress hormone and sometimes gets a bad reputation for contributing to belly fat. But cortisol plays a big role in immune health by modulating inflammation and immune cells. It's also responsible for regulating blood sugar and blood pressure, making it pivotal in the prevention of diabetes and heart disease.

DHEA (DEHYDROEPIANDROSTERONE)

DHEA is produced in the adrenal glands and is considered an antiaging hormone because it diminishes wrinkles, increases energy, enhances memory, reduces body fat, and improves libido. DHEA is converted to estrogen and testosterone, which is why we lean on our adrenal glands for hormone support after the ovaries stop producing hormones in menopause.

Non-Steroid Hormones

Non-steroid hormones are equally important and work in conjunction with steroid hormones to create that beautiful symphony we talked about. These hormones are not inside cell membranes like steroid hormones but instead are located on the surfaces of cells because they are water-soluble. Non-steroid hormones are made of amino acids and bind to receptor proteins on cell membranes, which then activate an enzyme inside the cell.

THYROID HORMONE

A butterfly-shaped gland located at the front of your neck called the thyroid is responsible for secreting thyroid hormones, namely thyroxine (T4) and triiodothyronine (T3). T4 is the inactive hormone and depends on other tissues like your gut, kidneys, and liver to activate it to T3. T3 is responsible for mood, energy, and metabolism, and when levels of T3 are too low, women experience irregular menses. We'll explore this hormone in depth in chapter 7, but overall low thyroid hormone is also associated with infertility, miscarriage, and digestive and skin disorders, including hair loss. I see a lot of women for hypothyroidism, and it can have a major impact on your sex hormones.

INSULIN

This hormone is produced by the pancreas to help bring blood sugar into your cells. Chronically elevated insulin can lead to higher levels of estrogen. In the case of PCOS, insulin and luteinizing hormone work together to stimulate the production of androgens like testosterone. In addition, insulin upregulates testosterone production in the adrenal glands, which we'll take a closer look at in chapter 8.

LEPTIN

Leptin is produced by fat cells and regulates your appetite by telling your brain about your hunger status. Women who have leptin resistance don't get the proper message to their brain and as a result can have difficulty losing weight.

OXYTOCIN

This hormone is often called the love hormone or cuddle hormone because it facilitates bonding—mother-and-baby bonding and bonding with your mate. It also reduces anxiety and opposes the harmful effects too much cortisol can have on you.

Have one of those friends whose memory is like an elephant's? She may just be pumping more oxytocin in her system since it's the hormone that helps solidify memories.

You don't need to have a baby or to breastfeed to bump up this hormone. A twenty-second hug can help elevate your oxytocin . . . or better yet, have an orgasm! (See chapter 10 for why orgasms are good for your health!)

Your Sex Hormones Are Followers

Your hormones are always talking and paying attention to one another. You know how your parents used to say, "If all your friends were jumping off a bridge, would you jump off a bridge too?" Well, that's exactly what your sex hormones do. They're just a bunch of little lemmings, always checking what the other hormones are doing. That's why it's essential to take a holistic perspective and to address the root cause of your symptoms. And that's why in this book we talk not only about hormones but also about the liver, gut, adrenals, thyroid, and metabolic system—because they all play a huge role in hormone health.

How Do I Know If I Have a Hormonal Imbalance?

When things in life are going smoothly, you tend not to even notice, right? But when major life drama occurs and shit hits the fan—your marriage is on the rocks, your job is sucking the life out of you, or someone you love has a health scare—boy, do you notice. You *feel* it. And it starts to consume all areas of your life and your overall happiness. Your hormones have a similar effect: when they are operating smoothly, you feel good and don't notice; when they are out of balance, you begin to have uncomfortable symptoms that you can't ignore because they are screaming for attention.

For example, if you have too much estrogen, also known as estrogen dominance, you might experience breast tenderness, weight gain, heavy periods, headaches, and mood swings. On the flip side, low estrogen can cause night sweats, insomnia, depression, vaginal dryness, and incontinence. If your progesterone is too low, you may struggle with anxiety, have heavy or irregular periods and mid-cycle spotting. These are all symptoms of a hormonal imbalance.

When your sex hormones are out of balance, it's usually just the tip of the iceberg. Review your Hormone Quiz at the end of chapter 1, and consider having your doctor test your hormones to further investigate your symptoms.

TEST YOUR OVERALL HORMONE HEALTH

- adiponectin (a hormone produced by fat cells that protects against diabetes and heart disease)
- DHEA-S
- DUTCH (dried urine test for comprehensive hormones) Complete
- fasting insulin
- 4-point salivary cortisol
- FSH, LH, estradiol (best tested on day 3 of your menstrual cycle)
- progesterone (best tested from day 19 to day 22 of a 28-day menstrual cycle)
- sex hormone–binding globulin (SHBG)
- thyroid panel: thyroid-stimulating hormone (TSH), total T4 and T3, free T4 and T3, reverse T3, anti-thyroperoxidase (anti-TPO), and anti-thyroglobulin antibodies
- total and free testosterone

Common Hormone Imbalances

What are some of the most common hormone imbalances and what do they look like? The following are the predominant imbalances I treat in my clinical practice and their corresponding symptoms:

- **Too much estrogen:** Estrogen dominance causes heavy, painful periods, fibroids, cysts on the ovaries, breast tenderness, and fibrocystic breasts, and is associated with increased risk of cancer.

- **Too little progesterone:** Common symptoms of low progesterone include infertility, anxiety, insomnia, uncontrollable crying, and irregular menstrual cycles (they are typically shorter).

- **Too much cortisol:** When your cortisol is in overdrive, you get that "wired and tired" feeling, where your brain is going a million miles a minute but your body is exhausted. This is also why women start to get belly fat (and why it gets a bad rap).

- **Too little cortisol:** Linked to "adrenal fatigue" or hypothalamic-pituitary-adrenal (HPA) axis dysregulation, low cortisol can leave you totally exhausted with a weakened immune system. You may get sick all the time or have wounds that don't heal. You can also experience chronic headaches, especially waking with headaches, and it's really hard to get out of bed in the morning.

- **Too little thyroid hormone:** Hypothyroidism is super common in my practice and can cause irregular periods and infertility. You may also feel fatigued and have dry skin, hair loss, and brain fog.

- **Too much testosterone:** This can happen with PCOS or because of post-pill androgen rebound, and the symptoms include oily skin and acne. You'll basically feel like a teenage boy and want to rage like one too. You can lose hair on your head or develop hair in unwanted areas, like your chin, chest, or abdomen.

- **Too little testosterone:** This is also something I treat that no one talks about. It causes you to lose muscle mass and your libido, but you'll also feel less motivated in life. It can look a lot like depression (which can confuse doctors), and you may find yourself crying all the time. It is really common on the pill.

And, hey, guess what? These hormones usually travel in a pack like a herd of girls going to the bathroom—they all want to hang out together! So odds are you won't have just one hormone out of balance; you'll have multiple hormones out of balance, which can make it difficult to understand what's going on in your body. The imbalances may manifest in several different ways too. That's why it's important to look at the overall story of your menstrual cycle and what else is happening in your body, so you can understand how all of these things tie together.

In the next chapter, we'll take a look at how these hormonal imbalances create many of the symptoms that manifest in post–birth control syndrome.

Key Takeaways:
The Lowdown on Your Hormones

- The two primary female sex hormones that regulate your cycle are estrogen (your va-va-voom hormone) and progesterone (your chill hormone); estrogen is higher during days 1 through 14, and progesterone highest days 15 through 28.

- You can only get pregnant five or six days in your cycle, because while your egg lives for only twenty-four hours, sperm can live in your uterus for five to six days.

- Day 1 of your cycle is the first day of menstruation.

- The birth control pill works by stopping your brain and ovaries from talking.

- Your steroid hormones include pregnenolone, estrogen, progesterone, testosterone, cortisol, and DHEA. Your non-steroid hormones include thyroid, insulin, leptin, and oxytocin. Remember, they're a bunch of followers, so if one starts to crash, the others will join in.

- A range of symptoms can indicate a hormonal imbalance, such as weight gain, heavy periods, headaches, mood swings, exhaustion, acne, night sweats, insomnia, or infertility.

- If you have one hormonal imbalance, there's a good chance you have multiple hormones out of balance and they will manifest in several ways. But, girl, I got you in part III.

CHAPTER 3

POST–BIRTH CONTROL SYNDROME

"Every time I quit the pill, my skin rebels and my periods go crazy," Em shared in her first visit. She had started the pill after her doctor told her it would "normalize her periods." And it partially worked; after eight years on the pill, she went from four days of changing a tampon every hour to three days of a hardly noticeable period, and her doctor gave her the okay to just "skip a month" altogether. Em was loving this newfound freedom, but she wasn't loving the fear of quitting the pill. She'd tried to quit twice before she came to my office, each time experiencing—you guessed it—heavy-as-heavy-can-be periods that were totally unpredictable. But she also experienced a whole new host of symptoms she'd never faced before: anger, irritability, acne, and nonstop headaches. And it is this new set of symptoms, paired with the return of her previous symptoms, that is known as post–birth control syndrome, the subject of this chapter and one of the most important things I'll cover in this book. Who wouldn't be fearful about quitting the pill with those symptoms?

Em was in my clinic because she wanted to make this the last time she'd be on and off the pill. Her doctor had told her there was no chance of her managing her symptoms without it, but Em had a strong belief in her body's ability to heal and find balance without the pill.

The good news? The symptoms of PBCS don't have to last forever, and once you learn how to identify the root cause of your symptoms, you can rebalance your hormones and start feeling better. In my practice, I have successfully treated many women with PBCS with a comprehensive protocol that has helped them enjoy more energy, better periods, and better moods, and I'm sharing it with you in this book so that you can do the same.

Why Wasn't I Warned About Post–Birth Control Syndrome?

"Post–birth control syndrome" is a term that refers to the collection of signs and symptoms that arise when you stop taking the pill. These can be symptoms you were suppressing with the pill, or they can be added side effects the pill created that your body is waking up to. What are these symptoms? Well, they can range from headaches, mood swings, anxiety and depression, and acne to a total loss of menstruation, pill-induced PCOS, infertility, hypothyroidism, leaky gut, and immune-related symptoms, which we'll get into in more detail in later chapters. These symptoms will generally appear within the first four to six months after discontinuing the pill.

Why does this occur? Because the basic mechanism of the pill is to flood your body with enough hormones that your brain stops communicating with your ovaries and you cease to ovulate. If the pill essentially shuts down the conversation between your ovaries and your brain, then it's no surprise that once you stop taking it you may encounter some challenges reestablishing the connection—not to mention the strain it has created on your adrenals, thyroid, gut, and liver. And that can have some long-term effects if you've taken the pill for a substantial amount of time.

Women who opted to take the pill to treat symptoms of PMS, heavy or irregular periods, acne, cyclical headaches, mood swings, or anything other than preventing pregnancy can expect a return of these symptoms if they are not supporting their body when they call it quits. Why? Because these symptoms all indicate a deeper imbalance that can be masked by the pill. True, the pill seems to *help* all these symptoms on the surface, but it isn't a root-cause solution, and is ultimately doing more harm than good. Depending on how long the pill has suppressed your symptoms, they can come rushing back with a vengeance once you stop taking it. This isn't your body revolting—**this is your body speaking up after having its voice stifled for a very long time. And this is where using the pill can create a dependency.**

In This Chapter

- How to know if you have PBCS

- What to do for a missing period

- Natural remedies for hormonal headaches

- Why you may have acne like a teenager and be losing hair like crazy

- How treating symptoms with the pill can mean big trouble for your health

It's a common myth that if a woman started the pill for contraceptive needs, she won't experience PBCS. Clinically, however, I have found that the majority of women coming off the pill struggle with PBCS, not just those who previously had symptoms. If you think you might be suffering from PBCS, I want you to know that your symptoms are not in your head. And research has shown that it can take as long as nine months to normalize your cycles after stopping the pill. If you fall into this camp of taking birth control for hormone symptoms, you will have PBCS if you quit the pill without implementing the recommendations in this book. By following the 30-Day Brighten Program, you can minimize this transition.

So why didn't your doctor warn you about PBCS? It's likely because they didn't know, and frankly, some of them just haven't been listening or taking your symptoms seriously.

Doctors have their own biases they bring into the clinic that affect the way they listen to and treat women. I can tell you, I am not the only doctor to have had a patient share that previous doctors told them their symptoms were "all in their head," only to be diagnosed with a serious illness after years of struggling.

Studies have shown that women are less likely to receive adequate medical care when it comes to chronic pain and heart attacks. Ask a woman with endometriosis how long her chronic pain, heavy bleeding, and gut symptoms were dismissed before she finally found a doctor who listened. The *New England Journal of Medicine* reported in a 2000 study that women were seven times

Quiz: Are You Suffering from PBCS?

If you've come off the pill and have noticed a variety of uncomfortable symptoms, check the boxes that apply to you:

- ☐ I haven't had a period (amenorrhea) for more than three months after stopping the pill.

- ☐ My periods are heavy and/or painful.

- ☐ I've been struggling to get pregnant.

- ☐ I've been diagnosed with hypothyroidism since starting or stopping the pill.

- ☐ I have terrible acne that won't go away.

- ☐ I've been getting migraines, especially right before I'm about to get my period.

- ☐ I have frequent headaches, especially around my period or cyclically.

- ☐ I've noticed recent hair loss.

- ☐ I feel depressed or have been diagnosed with depression since starting or stopping the pill.

- ☐ I have trouble with high blood sugar or hypoglycemia.

- ☐ I struggle with anxiety, nervousness, or worry since starting or stopping the pill.

- ☐ I feel gassy or bloated.

- ☐ I've noticed changes in my bowel movements.

- ☐ I suffer from inflammation and other immune imbalances.

If you have more than one of these symptoms, and especially if you have several of them, you are likely suffering from PBCS.

more likely to be discharged and misdiagnosed while having a heart attack. A heart attack!

I've been in health for over two decades now and can remember when people were called crazy for saying "leaky gut," which is now well accepted as a term for intestinal hyperpermeability. Doctors who prescribed probiotics a few decades ago were ridiculed for the practice, and now conventional medicine prescribes these regularly. I remember when people who said "adrenal fatigue" were called quacks, and now HPA dysregulation is a documented phenomenon in the literature. And you better believe that PBCS is no exception.

We need a whole lot more humility in medicine, and recognition that we don't know everything, especially since we just started including women more regularly in studies like yesterday. Your doctor should be curious, not dogmatic, about your health. If they don't seem concerned about the symptoms you bravely express to them in your appointments, and aren't doing the follow-up research to try to understand what's going on in your body, my best suggestion is to find a new doctor. There's someone out there who will do better, believe me.

The issues surrounding medical gender bias and how it hurts women are becoming well recognized in the research, and medical schools are working to change medical education. Sadly, it's going to take a long time for medicine to change, which is why I'm bringing the medicine directly to you in this book.

Where Did My Period Go . . . and Is It Ever Coming Back?

Have you been off the pill for three months with absolutely no sign of your period? Is a part of you wondering if it's *ever* coming back? For many women, coming off the pill means the end of regular periods—especially if they began taking it because they had irregular periods. In fact, some women will never see it return.

Studies have revealed that women who stop the pill can experience major changes to their menstrual cycle, including shorter luteal phases, longer cycle lengths, and anovulatory cycles (cycles in which no egg is released) for months—and sometimes years. One study found that the average time for anovulatory cycles was about nine months. Other studies have shown it can take longer than expected to get pregnant, even years.

Losing your period altogether, also known as post-pill amenorrhea, can last for four to six months after stopping the pill, sometimes longer without intervention. If you're worried that your period has gone missing for too long, it's time to go searching for it. The first step is to have testing done so you can begin to understand why your period hasn't returned and what you need to do to restore your hormone health.

Sadly, many women quit the pill with baby making on the agenda, only to find themselves unable to conceive—faced with missing periods, infertility, and multiple miscarriages. This is a big reason I got into the work that I do and began digging into the research regarding the pill's impact on our fertility. Like many doctors, I was taught initially that the pill had little impact on fertility. It was only once I entered my advanced fertility curriculum that my clinical instructors began to share that women should come off of hormonal contraceptives years before they want to become pregnant because there may be a significant delay.

As we will explore in part II of this book, the pill depletes nutrients, raises inflammation, stresses the adrenals and thyroid, causes insulin dysregulation, and disrupts the microbiome—any of which can have a detrimental effect on fertility. It makes sense we would need to give our bodies some major TLC before conception. But as we'll discuss in chapter 10, there is more to the story than what we've been told, and you'd better believe I've got some solutions for you. The truth is, there are a whole lotta skeptics in this arena and a deficit of substantial research. Some of the most painful conversations I've had in medicine involve telling women who want nothing more than to hold a baby in their arms that their labs and symptoms all point toward infertility. But these have also led to some of my most gratifying cases, when we restore ovulation and optimize conception, and I get that call of "I'm pregnant!" after we employ the protocols I will teach you in this book.

SOLUTIONS FOR RECOVERING YOUR MENSTRUAL CYCLE

If your period has gone missing, there are a number of steps you can take right now to begin reversing the effects of the pill and restore your menstrual cycle. Your liver has been working overtime to process all the synthetic hormones in the pill so they can be moved out by your gut. One of the first steps to recovering your body and cycle is to support your liver. We'll go into more depth about how to do this in chapter 5, and the 30-Day Brighten Program includes plenty of liver-loving food.

The birth control pill also depletes a lot of nutrients crucial for hormone balance and health. In addition to helping with PMS and other PBCS symptoms, and because they work so closely with your hormones, nutrients play an

Six Tests for a Missing Period

If you've been off the pill for three months or more and have not gotten your period, test your hormones:

1. **Pregnancy test.**

2. **FSH, LH, estradiol:** FSH and LH are brain hormones that tell you how your brain is talking to your ovaries. A test of estradiol will help you understand how your ovaries are responding to signals from that hormone. An elevated FSH and low estradiol are a sign your ovaries are not functioning optimally and may be due to primary ovarian insufficiency (POI).

3. **Prolactin:** Prolactin blocks you from having your period by inhibiting the secretion of FSH. The same mechanism that causes TSH to go up can also elevate prolactin levels. It could be correlated with what's going on in the thyroid or due to a condition called prolactinoma, a benign brain tumor.

4. **Thyroid panel** (TSH, total T4, total T3, free T4, free T3, reverse T3, anti-TPO, and anti-thyroglobulin antibodies): If your thyroid isn't functioning properly or you're hypothyroid or hyperthyroid, it can affect your menstrual cycle, including whether or not you ovulate or get a period.

5. **Total and free testosterone:** Elevations in testosterone can suppress ovulation and stop your period.

6. **Adrenals** (4-point salivary cortisol or 4-point urinary cortisol and cortisone with DHEA-S): Your adrenals produce cortisol, especially in response to stress, which can decrease progesterone and leaves estrogen unopposed.

---

essential part in recovering your missing period. Key nutrients you'll want to replenish include vitamin B6, magnesium, and zinc, all of which are depleted by the pill. Taking a prenatal or multivitamin can often help meet these needs.

In certain cases of amenorrhea when prolactin levels (the hormone associated with breast milk production) are really high, studies have shown that vitamin

B6 can lower those levels and restore a period. While prolactin is typically associated with women who are pregnant or have recently given birth, a high amount of prolactin can be due to other reasons, which should be investigated. We know too that B6 is typically lower in women who have PBCS, which is why it's an important nutrient in PBCS therapy. (An added bonus: B6 has been found to be beneficial in both preventing and alleviating symptoms of PMS.)

Vitamin B6 is also essential in the development of the corpus luteum (what remains after an egg is released and is responsible for secreting progesterone). To reestablish ovulation and have a regular menstrual cycle, make sure you're taking care of the corpus luteum, which helps with fertility too. A healthy corpus luteum will help ensure you have ample levels of progesterone as well; when progesterone dips, you may feel irritable and cranky and not in control of your mood. If you have signs of too little progesterone, incorporate foods into your diet that are rich in B6, like wild-caught fish, grass-fed beef, chicken, sweet potato, spinach, and banana. Especially if you're still on the pill, you'll want to start bringing in these foods and fixing your gut as soon as possible (there will be more on your gut in chapter 6).

Vitamin B6 is utilized by more than one hundred enzymes involved in your metabolism, so it's essential that you properly supplement with it. In a systematic review and meta-analysis of nine randomized placebo-controlled trials, it was suggested that up to 100 milligrams per day may be the most beneficial B6 dose for treating PMS and mood-related symptoms. There's still not enough evidence to determine if women should be taking a dose that high, but it's certainly something worth discussing with your doctor, especially if you have PMS, symptoms of PBCS, or post-pill amenorrhea.

Another extremely important nutrient is magnesium. If you're on the pill or experiencing any of the symptoms of PBCS, then you must start taking magnesium, which helps thousands of processes in the body—I can't even begin to cover all of them in this book. In terms of which relate to hormonal health, magnesium is key in controlling insulin production, which influences your testosterone levels and the health of your ovaries overall. Remember, good blood sugar control is necessary for optimal adrenal health, and that's the foundation of balancing hormones. Magnesium has the added benefit of reducing sugar cravings.

Magnesium is crucial for phase II detoxification in the liver too, which is how you move estrogen out of your body, helping to promote balance between estrogen and progesterone and make your tissues more likely to respond to the right hormones at the right time. Magnesium is found in fish, spinach, almonds, and molasses, but because our food supply isn't as rich in magnesium as it once was, I recommend taking a magnesium supplement of 300 to 600 milligrams daily. Choose magnesium citrate if you struggle with difficult-to-pass stools or constipation; otherwise look for chelated magnesium like magnesium bisglycinate (see Resources on page 333).

Zinc is another critical mineral for women with PBCS. Although it's found in foods like pumpkin seeds, red meat, chicken, oysters, clams, and lots of other shellfish, it can be hard for your body to replenish its stores in the short term. (Unfortunately, the zinc you might find in whole grains, legumes, and nuts is not as bioavailable because it is bound to phytic acid.) And in the short term, any kind of zinc deficiency will cause issues with your hormones overall and make it hard to recover your period. Any time you take zinc, partner it with copper to avoid a copper deficiency. Aim for 15 to 30 milligrams of zinc per day, depending on how much of it you are also getting through food sources.

In addition to these supplements, make sure you eat a nutrient-dense diet, with enough fat and calories. This provides your body with good blood sugar stabilization, which will tell your adrenals the environment is safe and make it more likely you will be able to regulate your period again. Healthy fats fuel healthy hormones and keep inflammation low. Following the 30-Day Brighten Program in chapter 12 will get you started.

When you're trying to recover a missing period, it's also important to support your body's circadian rhythm, because if you're not sleeping, then you stand zero chance of getting your hormones back on track. Both quantity and quality of sleep are vital. See chapter 11 for guidelines on how to reset your circadian rhythm (page 233).

Why Is My Period So Heavy, Painful, and _____ ?

Maybe a missing period isn't your problem, but instead your period is so bad that you almost wish it *would* go missing. This is another common symptom of PBCS, because many women start taking the pill to get rid of heavy or painful periods, and when they discontinue the pill, typically within about two to four months those periods can return more forcefully than ever.

The underlying cause of many period problems is—you guessed it—a hormonal imbalance, most commonly estrogen dominance. Diets rich in sugar, refined carbohydrates, alcohol, non-organic meat, conventional dairy products,

How Poor Sleep Affects Your Hormones

- Poor sleep habits, insomnia, and sleep deprivation have been shown to cause changes in key hormones that affect the menstrual cycle.
- Elevations of follicle-stimulating hormone (FSH) have been found in night workers and those who sleep during the day and are up at night. High levels of FSH are a sign of poor ovarian reserve and possibly advanced ovarian aging.
- Luteinizing hormone (LH) levels are higher in people who skimp on their sleep, which is associated with infertility.
- A poor or skipped night of sleep can cause a rise in thyroid-stimulating hormone (TSH), which can stop you from ovulating.
- Estradiol levels can elevate with poor sleep, which can affect your ability to ovulate and regain a regular cycle.

and caffeine can all exacerbate this imbalance. Stress, environmental pollutants, and hormone-disrupting chemicals also contribute. In chapter 4, I'll help you decode what your period is trying to tell you, with some quick solutions that can alleviate your uncomfortable symptoms right away. If your painful periods are seriously wrecking your life, then skip ahead to page 67 for some pain-relieving remedies and be sure to meet with your doctor to find out why.

Hormonal Headaches

Some women start the pill to get rid of hormonal headaches. Others develop hormonal headaches because they are *on* the pill. Clinically, I have seen many women suffering from headaches as part of PBCS.

Hormonal headaches are more than just a monthly inconvenience or burden—*they are a sign that something is out of balance in your body and in need of immediate attention.* Hormonal headaches have a root cause, which is most commonly estrogen dominance but can also be an imbalance of other hormones, like thyroid or cortisol, as well as nutrient deficiencies, like magnesium and B2.

While you look for the root cause, it's important to have some go-to remedies to alleviate pain. If you find yourself reaching for NSAIDs like ibuprofen when you get hormonal headaches, **step away from the medicine cabinet**. When you take pain relievers regularly, they can suppress ovulation and make it more difficult to recover from hormonal imbalances. They can also lead to more frequent and more painful headaches—exactly what you're trying to avoid when you pop one or two of these pills on a regular basis.

Here are some natural remedies for those annoying (and seriously painful) hormonal headaches:

- **Vitamin B2 (riboflavin)** at a dose of 400 milligrams a day helps many people to reduce the number of migraines they have. Riboflavin is more of a preventative therapy and has to be taken consistently for at least one month to see any effect; three months is the ideal minimum amount of time to evaluate the therapy.

- **Magnesium** has been shown to prevent headaches. It acts as a muscle relaxant and is also anti-inflammatory. Aim for 300 to 600 milligrams of magnesium bisglycinate nightly.

- **Bromelain** is derived from the core of a pineapple and is a natural way to break down the inflammation-causing molecules in your body. When taken with food, it acts as a digestive enzyme. But taking about 200 to 300 milligrams twice daily between meals can help lower your pain and inflammation.

- **Feverfew** has been shown to prevent migraines. I recommend women aim for at least 25 milligrams daily to get the most benefit from this anti-inflammatory herb.

- **Turmeric** is a beautiful golden root that works on some of the major inflammatory pathways in the body to reduce pain and inflammation. It's excellent as a beverage (see the Anti-inflammatory Turmeric Spritzer, page 304) or can be taken in capsule form at a dose of 1,000 to 2,500 milligrams daily.

- **Ginger** rivals NSAIDs and has been shown to be as effective at reducing pain. As a supplement, a dose of 1,000 milligrams twice daily works well for most people. Ginger is also lovely as a tea and can be combined with turmeric for double the herbal anti-inflammatory power!

- **Essential fatty acids**, especially omega-3, are found most commonly in fish oils. They lower inflammation and can help bring hormones back into balance. Aim to eat two to three servings of fatty fish or supplement with 1,000 to 2,000 milligrams daily. Yes, ladies, sardines can help your headaches!

- **Hydration**—I know it seems obvious, but for reals, if you're dehydrated and prone to headaches, then odds are one is brewing. Dehydration is a common cause of headaches. Aim for a minimum of 80 ounces of water, herbal tea, bone broth, or mineral water daily.

- **Essential oils**, particularly one or two drops of peppermint or lavender oil, applied or massaged into the temples can safely alleviate headaches.

- **Movement** such as stretching and exercise can go a long way toward keeping pain at bay. Consider working with an exercise physiologist, functional trainer, physical therapist, or other movement expert as part of your pain-prevention regimen.

- **Massage** cannot be underestimated for women who suffer from hormonal headaches. Bodywork will not only soothe your stress but also allow your nervous system to take some much-needed relaxation time while your muscles get a good release.

My Acne Is So Bad I Feel Like a Teenager!

Because the pill reduces testosterone, it also lowers oil production, which is why it can be effective in eliminating acne. But the pill doesn't fix acne; it simply suppresses it—if you're lucky (not all women see better skin with the pill). So when you do decide to come off the pill, you will likely encounter a rebound of testosterone and a flare-up in your acne.

I hate to say it, but if you're struggling with acne because of PBCS, your skin may get worse before it gets better. Think about it: your skin is not your most vital organ, so your body is going to heal your more important organs first (like the ones that keep blood and oxygen flowing). Also, the pill creates gut inflammation, suppresses healthy hormones, and burdens the liver so it can't clear out unnecessary hormones. This is why so many women have horrendous acne return when they stop their birth control pills.

I gave you the bad news first, so now I'll share the good news. In my clinical experience, most women who are actively working to heal their body experience a milder acne flare for the first couple of months post-pill, but it generally decreases by months three to five. In women *not* actively working to undo the effects of the pill, this skin battle is often much longer. Since you're reading this book, you're in the former group and will begin to see improvement once you implement the strategies in the 30-Day Brighten Program and eliminate the chemicals that are messing with your hormones.

Here are some strategies to get you started:

- **Heal your gut.** Your skin can be a reflection of what's going on inside and often points to inflammation in your gut. Read chapter 6 and begin the 30-Day Brighten Program, which includes a whole-foods, anti-inflammatory diet.

- **Eat healthy fats.** Healthy fats like olive oil, macadamia nut oil, cold-water fish, and avocados are better for your skin than inflammatory fats like canola or corn oil.

- **Consume probiotic-rich foods.** Good gut bugs like those in probiotics can have a dramatic impact on your skin health. Start eating more fermented foods, such as fermented beets, kefir, kimchi, kombucha, and sauerkraut. You can also start taking a probiotic supplement (see chapter 6 for more information).

- **Eliminate dairy.** Dairy is often the culprit when it comes to acne. Cut it out for at least six weeks and see if you notice any improvement.

- **Clean up your cosmetics.** Many beauty products contain toxic chemicals that mess with our hormones, which can also make acne worse. Swap out these products for cleaner alternatives (visit www.ewg.org/skindeep, and see chapter 5).

- **Increase dietary zinc.** Zinc can help your body eliminate extra testosterone, which is often the underlying cause of skin conditions like acne. Zinc is found in beets, carrots, egg yolks, oysters and other shellfish, pecans, pumpkin seeds, and red meat.

- **Try saw palmetto.** This herb can prevent testosterone from converting into DHT (dihydrotestosterone), which causes acne (typically around the jawline) and hair loss.

- **Include Vitex.** Chaste tree berry, or Vitex, is an herbal supplement that has been shown to help balance hormone levels and address symptoms like cysts and acne.

- **Get enough omega-3.** This essential fatty acid decreases inflammation and supplies your skin with healthy oils.

- **Eat foods with GLA.** Gamma-linolenic acid reduces prostaglandins, which can combat acne and cramps.

- **Try licorice root.** This root can reduce inflammation, oily skin, and acne. (Caution: If you have high blood pressure, do not take licorice.) Aim for 20 milligrams daily in supplement form.

- **Cook with turmeric.** Turmeric can reduce inflammation and support detoxification of the liver.

- **Do a post–birth control detox.** See chapter 5 (page 85) to complete this detox.

WTF, Why Is My Hair Falling Out?!

Are you waking up with a pillow covered in hair or do you have a clogged shower drain? Maybe you're noticing your part getting wider? This is yet another common symptom that rears its head during PBCS. Hair loss is a sign of a deeper hormonal imbalance, and it's important to seek treatment immediately if you've begun to feel like you're shedding hair everywhere.

A number of tests can help you identify the root cause of your hair loss:

- **Complete blood count (CBC) and ferritin blood test:** Both tests screen for iron deficiency. When your iron dips, hair loss can occur. If your ferritin is below 50 ng/mL, inadequate iron may be the culprit.

- **Thyroid panel:** TSH, total T4 and T3, free T4 and T3, reverse T3, anti-thyroperoxidase (anti-TPO), and anti-thyroglobulin antibodies can all help pinpoint if your thyroid is to blame.

- **Antinuclear antibodies (ANA), celiac panel, or Cyrex Array 5:** These tests screen for autoimmunity—there are many autoimmune conditions that reveal themselves with hair loss.

- **High-sensitivity C-reactive protein (hs-CRP) and erythrocyte sedimentation rate (ESR):** Both of these check for inflammation, which is a sign that you may be at risk of developing other symptoms or disease. Clinically, I prefer hs-CRP.

- **Total and free testosterone, and dihydrotestosterone (DHT):** Most commonly associated with male-pattern baldness, testosterone can also cause women to lose their hair. Elevated testosterone may be due to hormonal imbalance or PCOS.

- **FSH, LH, estrogen, and progesterone:** Imbalances in these hormones will cause not only hair loss but also PMS, irregular periods, and/or heavy periods.

- **Sex hormone–binding globulin (SHBG):** This protein grabs onto excess hormones. If it's low, then your free testosterone may be elevated as a result.

- **Cortisol and DHEA sulfate (DHEA-S):** Think of cortisol as an aging hormone. If cortisol is high or DHEA-S is low, you will visibly age quicker, which includes hair loss.

- **Comprehensive metabolic panel (CMP):** This is a good baseline to check liver and kidney function as well as the balance of your electrolytes.

- **Fasting insulin, fasting blood glucose, and hemoglobin A1C:** These labs screen for diabetes. Improper blood sugar regulation can cause imbalances in your hormones and impede circulation to your scalp.

- **Heavy metal test:** Are you an artist or welder, do you have a lot of silver fillings, or do you eat a lot of rice or non-organic foods? If so, you may have heavy metals driving your hair loss.

What to Do About Hair Loss

Who would've thought there could be so many different causes of hair loss, right? Fortunately, there are many natural remedies and supplements available to stop your hair loss and regrow those luscious locks once again.

- **Iron** may be required, in your diet or as a supplement, if you have low ferritin or iron deficiency anemia.

- **B complex** vitamins are involved in a ton of metabolic pathways in your body. Take a quality B complex that contains B12 and folate (not folic acid).

- **Adaptogenic herbs** will help regulate your cortisol and begin to balance your hormones overall. My favorites are Rhodiola, ashwagandha, and licorice root. (Caution: If you have high blood pressure, do not take licorice.)

- **Saw palmetto**, as I mentioned earlier, helps prevent the conversion of testosterone to DHT, which causes hair loss.

- **Apple cider vinegar**, 1 to 2 teaspoons taken before meals, will help raise your stomach acid, which will allow you to absorb more vitamins and minerals.

- **Pumpkin seeds** are a great source of zinc and also help prevent the conversion of testosterone to DHT. See page 258 for how to use them in seed cycling.

- **Essential fatty acids**, as I already mentioned, can be found in salmon or sardines, and help reduce inflammation. Having fatty fish at least once per week is one way to increase your omega-3s. Chia seeds and walnuts are also a great source.

- **Freshly ground flaxseed** increases SHBG, which binds excess hormones. I recommend 2 tablespoons daily. The seeds must be freshly ground (pre-ground flaxseed—flax meal—is often rancid before it even hits the grocery store shelf).

- **Exercise** increases your circulation, reduces your stress, and balances your blood sugar all at once. If you feel stressed out, consider yoga or Pilates. If you have blood sugar or hormonal imbalances, you may need more strength training.

What Your Hair Loss Pattern Tells You About Your Hormones

Patchy	A cortisol imbalance, a deficiency in B vitamins or zinc, or a heavy metal exposure
Thinning	A hormonal imbalance, such as thyroid
At the top of the head	Testosterone, progesterone, cortisol, or estrogen out of balance
Balding all over	Poor circulation or a deficiency in protein, essential fatty acids, B vitamins, silicon, or zinc
Total body hair loss	DHEA, blood sugar regulation, or circulation issues

- **Laser therapy,** as clinical trials have shown, can prevent the progression of hair loss and regrow thinning hair. In my clinic, I recommend the Capillus laser therapy cap for patients because it's easy and convenient to use and has been FDA-cleared in the treatment of androgenic hair loss.

The 30-Day Brighten Program to Reverse PBCS

You now have the CliffsNotes version of how the pill takes a serious toll on your body and the many symptoms of PBCS. So, what next? In chapter 12, the 30-Day Brighten Program will assist you in rebalancing your hormones to help you recover from PBCS. Whether you're still taking the pill for birth control reasons or have recently come off it and are experiencing the symptoms of PBCS, the program will help reduce the uncomfortable side effects. If you're coming off the pill, you may go on a bit of an emotional roller coaster, which we'll address in chapter 9.

But first, let's take a deeper look at the major systems in the body that are affected by the pill, such as your liver, gut and immune system, thyroid and adrenals, metabolic system, mood, and even libido and fertility. For each of these systems, I offer specific tools for reversing the harmful effects of the pill, so if you need to focus on a particular area, like your gut or your thyroid, you can start your healing process right away.

What Happened to Male Birth Control?

A 2016 trial of injectable male contraceptive was suspended after determining that the risks to men's health outweighed the benefits. What brought researchers to that conclusion? Out of 320 men, 20 said they couldn't tolerate the side effects, which included pain at the injection site, acne, and depression. Yes, based on *20 men* saying they were uncomfortable with these symptoms, researchers discontinued the trial despite it being 96 percent effective. For the record, in early birth control trials in women, 15 percent of women had side effects and three women died . . . without investigation.

And these side effects? They are the exact same symptoms women experience at a higher rate when taking the pill and other forms of hormonal birth control. Well, not the exact same symptoms. While most women experience a low or absent libido using oral contraceptives, these men experienced an *increase* in libido.

Many women—about 100 million worldwide—opt for hormonal contraceptives, which deliver the same side effects *and more* compared to the male version yet are deemed completely acceptable for women. Side effects women have dealt with since the FDA approved the first birth control pill. Side effects over which women have voiced concern. Side effects that are often dismissed. Now, I'm not about to advocate that we start giving hormones to the men in our lives, but I do think we should be asking why we aren't improving our options for women.

Key Takeaways:
Post–Birth Control Syndrome

- Almost 60 percent of women take the birth control pill for reasons other than pregnancy prevention.

- Post–birth control syndrome is a term that refers to the collection of signs and symptoms women develop when they stop hormonal birth control.

- Common symptoms of PBCS include loss of your period, a heavy or painful period, acne, headaches, hair loss, depression, and anxiety.

- If your period has gone missing, test your hormone levels: FSH, LH, estradiol, thyroid, prolactin, adrenals, and total and free testosterone.

- Vitamin B2, magnesium, bromelain, feverfew, turmeric, ginger, essential fatty acids, hydration, and peppermint or lavender oil can all help alleviate hormonal headaches.

- Stop hair loss with iron, B vitamins, adaptogenic herbs, saw palmetto, pumpkin seeds, essential fatty acids, freshly ground flaxseed, and exercise.

- The 30-Day Brighten Program can help you rebalance your hormones and reverse the symptoms of PBCS.

CHAPTER 4

TAKE BACK YOUR PERIOD

You don't need the pill to fix your period. In fact, the pill can't fix your periods, your acne, your PCOS, your PMS, or your hormones, because it is designed to shut down your entire reproductive system and psych your body out. The reality is that the only person who can fix your period is you, with the help of your doctor. Those symptoms that you hate (or make you think your body is hating on you) are exactly what you need to figure out your period problems and make them gone like yesterday. Every day in my medical practice I meet with women who were told that the pill is *the* solution and in this chapter I'm calling BS. You don't have to rely on the pill as a symptom solution, and I'm going to offer a whole lotta alternatives. And look, if you're the one woman in a million who doesn't have period problem complaints, then you can skip to the next chapter. But if you're like the majority of my patients, then know that I've got you.

Your period and menstrual cycle supply you with some serious data about your health. In fact, what your period says about your health is so important that the American College of Obstetricians and Gynecologists decided that it was a new vital sign. That's right, the fifth vital sign. Your period is up there with temperature, pulse, respiratory rate, and blood pressure. You know, all the things that tell your doctor you're alive and are used to assess your health. So, what happens when you take the pill? Well, you rob yourself and your doctor of valuable data that can help you understand both what's happening currently in your body and what's coming down the pipeline if you don't intervene.

In this chapter, I'm going to help you interpret your fifth vital sign and understand what it means if your periods are heavy, light, long, short, painful, irregular, or missing altogether. If you're thinking about using the pill to avoid dealing with these, I'd like to offer you an alternative solution. Changes in your cycle can indicate anemia, thyroid disease, endometriosis, stress, and much more. Let's pause for a moment and recognize that if your doctor put you on the pill to treat your period symptoms without any further investigation, then you could be finding yourself in big trouble. Take a breath. If you've already been using the pill to eliminate unwanted symptoms, know that I've got you and that this chapter is going to help you next-level your period knowledge.

When you learn to decode what your period is telling you, you gain a vast amount of knowledge about not only your overall health but also your fertility. What's going on in your cycle directly affects your fertility and can often help you to discover why you may be struggling to conceive. Now that you understand your cycle and why it's so badass, let's explore what that fifth vital sign is telling you.

Interpreting Your Fifth Vital Sign

In chapter 3 we covered many of the symptoms of post–birth control syndrome, some of which had to do with your period, and now we're going to decipher your symptoms and what they indicate about your health. With a "normal" period, you may have minor increases in emotion, some fullness, or light cramping, but for the most part your period should sneak up on you symptom-wise. It should *not* derail your life the way some of the symptoms we'll talk about in this chapter do. In this section, we'll take a deep dive into the different types of periods you may experience and what they signify. For example, a heavy period may be a sign of fibroids, whereas a light period generally suggests low estrogen, which could stem from a low-fat diet. And while nobody likes period problems, I'm going to teach you how to understand what those symptoms mean and the common ways your body communicates through your period.

Very Heavy or Long Periods

Let's be clear about something: needing to change a tampon or pad every hour, seeing giant clots in your menstrual blood, or feeling like your period goes on

---

In This Chapter

- Decoding your fifth vital sign and what your period says about your health
- How to know if your period is too heavy
- What painful periods really mean
- Why you may have mid-cycle spotting or pain with sex
- Natural remedies that are better than popping ibuprofen

---

indefinitely is not a normal state of being, and it's not a normal part of having a period. I know what it's like to be the girl who has her period for seven to eight days and carries a change of clothes with her and a backup sweatshirt to wrap around her waist just in case one of her tampons or pads fails. Common causes of heavy or long periods (more than seven days) include iron deficiency anemia, estrogen dominance, thyroid disease, fibroids or polyps, endometriosis, and certain cancers. In my practice, the most common reason I see for heavy or long periods is estrogen dominance and thyroid disease. We'll explore more about what you can do for estrogen dominance a bit later in the chapter and how to optimize your thyroid health in chapter 7.

The Paragard, a copper IUD, is commonly associated with heavy or long periods, which is why it's not advised for women who already experience these symptoms. If you have the copper IUD, meet with your doctor to determine if it's the right contraception for you. (See chapter 13 to evaluate your options.)

LAB TESTING FOR HEAVY PERIODS

Hormone testing is a must with this amount of bleeding. Having a complete thyroid panel, along with estrogen and progesterone testing, can help determine if there is an imbalance contributing to your symptoms. You also want to evaluate if you have iron deficiency anemia. I recommend the following tests—on the noted day of your cycle for those tests that are day-specific:

- day 3: estradiol
- days 19 to 22: serum progesterone and estradiol or DUTCH test

- thyroid panel (TSH, total T4 and T3, free T4 and T3, reverse T3, anti-TPO, anti-thyroglobulin antibodies)

- CBC

- ferritin

WHAT TO DO RIGHT NOW

While your lab testing is being completed, you can begin making powerful diet and lifestyle changes to support a healthy menstrual flow and increase your energy. Incorporate iron-rich foods into your diet, such as red meat, leafy green vegetables, blackstrap molasses, and, if you can hang with it, liver. Eating foods rich in B12, B6, and folate can help with red blood cell production and boost your energy. Use the 30-Day Brighten Program meal plans in chapter 12 to choose foods that will support your estrogen metabolism and replenish your nutrient stores. Be sure to read the section on estrogen dominance later in the chapter, as this may apply to you. If you suspect your thyroid may be an issue, read chapter 7 and begin the recommendations outlined there.

Menorrhagia (aka Extremely Heavy Periods) Checklist: How to Know If Your Period Is Too Heavy

If you check even one box of these, please see the related "Lab Testing" section (page 63) for evaluation options.

- ☐ I change a pad or large/super tampon every hour for three or more hours in a single day.
- ☐ I fill a menstrual cup three times in a day.
- ☐ I need to double up on menstrual products to control my flow.
- ☐ I wake to change menstrual products during the night.
- ☐ I limit my activities due to menstrual flow.
- ☐ My period lasts longer than seven days.
- ☐ I have blood clots that are larger than the size of a quarter.

Menstrual Blood Stem Cells

Did you know your menstrual blood is so incredibly powerful that it can aid in stroke recovery, improve liver function in cases of liver failure, and help lungs heal after injury? Menstrual blood is rich in stem cells, which are showing a lot of promise in animal studies. Your period is truly powerful. It kind of makes you wonder how all that period-shaming business ever got started. Don't let anyone tell you your period isn't awesome—your menstrual blood has superpowers! It has the ability to heal and regrow damaged tissues that could otherwise cost someone their life.

Painful Periods

Are you clenching a bottle of Midol every month during your period? Or maybe you call off the gym and all social activities because you know period pain is about to take you down. Intense menstrual cramps, also known as dysmenorrhea, usually begin just before menses, although they can start a few days before, and last for about three days on average. This pain can be localized to your uterus or it can radiate to your low back as well. Some women experience nausea, vomiting, headaches, and diarrhea with it too.

Painful periods can be a sign of infection, endometriosis, fibroids, or ovarian cysts, but from the conventional perspective, pain is incorrectly considered a normal part of menstruation and many concerned women are often dismissed. Research has shown that the rate and duration of pill use for severe menstrual cramps during adolescence is higher in women who are later diagnosed with endometriosis. This is one more condition the pill can be masking! It's also important to note that adhesions—scar tissue that is common in endometriosis—can be responsible for an estimated 12 percent of female infertility cases.

So just what causes these painful periods? Prostaglandins. These hormone-like chemicals stimulate contraction of the uterus, and when there is too much contraction, there is pain. While the birth control pill may offer symptom relief, it doesn't address the underlying cause, which is ultimately inflammation. In fact, while the pill is suppressing the symptoms in your uterus, it is creating inflammation in other areas of the body, which we'll discuss further when we address rebuilding the gut and handling metabolic mayhem.

LAB TESTING FOR PAINFUL PERIODS

If painful periods are a recent development, have a complete blood count (CBC) test to rule out infection and a human chorionic gonadotropin (hCG) test to rule out ectopic pregnancy. Your doctor may consider a transvaginal ultrasound to understand the cause of your pain. In my office, we investigate for underlying causes of inflammation, like gut or chronic infections, adrenal dysregulation, nutrient imbalances, and other stressors. Other helpful tests in understanding your root cause include:

- adrenals: DUTCH adrenal or 4-point salivary cortisol or 4-point urinary cortisol and cortisone with DHEA-S

- DUTCH Complete: On days 19–22 of your menstrual cycle, get the DUTCH test. The DUTCH Complete includes adrenal hormones.

- estradiol and progesterone: On days 19–22 of your menstrual cycle, have your estradiol and progesterone tested.

- fatty acids

- high-sensitivity C-reactive protein (a marker of inflammation)

- RBC magnesium

- stool culture, lactulose breath test, or other gut test (see chapter 6)

Why Does My Period Make Me Poop?

Ever wonder why you poop so much before your period? It's the shift in hormones that alter gut motility. In a naturally menstruating female, progesterone will rise after ovulation. Progesterone slows motility and can cause constipation when elevated (which is why early in pregnancy it can be hard to poop). Just prior to your period, your progesterone levels drop, which can allow for an increase in transit time—or, in other words, make you poop more.

Prostaglandins, the hormone-like molecules that stimulate the contraction of your uterus to shed its endometrium lining, also stimulate bowel contractions, resulting in more frequent bowel movements leading up to your period.

WHAT TO DO RIGHT NOW

If you suspect you have an underlying issue, then it's tim
doctor for a pelvic exam, lab testing, and possibly lab imaging, ...
nal ultrasound. Start eating a low-inflammatory diet by using the meal p...
chapter 12. It generally takes about two to three months for this diet to really
take effect if your period pain is significant, which is why I recommend using
the following supplements as part of the 30-Day Brighten Program to reduce
inflammation, balance your hormones, and eliminate period pain:

- **Magnesium glycinate**—I generally recommend 300 milligrams, twice daily five days before menses and for the first three days during. The rest of the month, aim for 300 milligrams nightly.

- **Magnesium oil** can be massaged on your abdomen the night before you expect menstrual cramping, and used for the duration of your cramps.

- **Cramp bark**'s name is no coincidence. It's an effective herb when it comes to menstrual cramps. I typically recommend taking 1 teaspoon of tincture two to three times daily, beginning two days before you expect your period and through the duration of your cramps.

- **Omega-3 fatty acid** at an approximate dose of 2,000 milligrams daily achieves the anti-inflammatory effect. The women in my clinic typically report needing to be on this for about two cycles to experience the benefits. There are several studies that support the use of fish oil for reducing menstrual cramps.

- **Vitamin E** reduces the number of prostaglandins your body makes. At 400 IU daily, it has been shown to be effective for cramps.

- **Vitamin B1 (thiamine)** taken daily at 100 milligrams has been shown to help with both muscle cramps and fatigue, side effects of low thiamine. I recommend taking this as part of a B complex.

In addition to supplements, applying heat to your abdomen can relieve cramps. Try using a hot water bottle over the lower abdomen. Just be sure not to leave it on for too long (more than twenty minutes), as this can cause issues for your circulation. You may also want to try acupressure, which can help with not only your cramps but also any lower back pain you may experience during your menstrual cycle.

Light Menses

I used to have this girlfriend who always bragged about her periods lasting only a couple of days and how she could get away with just a panty liner alone. And while an easy, breezy period may sound ah-mazing, this is actually a sign that your hormones are not balanced. If your period lasts less than three days, or you require little more than a panty liner during your period, then you may not have enough estrogen. If your estrogen is low, you may also be experiencing vaginal dryness, sleep disturbance, infertility, and even drooping breasts.

Low estrogen is a normal change in postmenopausal women, but if you're nowhere near menopause, then it's time for some root-cause investigation and a visit to your doctor. Common causes of low estrogen include eating a low-fat or vegetarian diet, overexercising, having a low body weight, primary ovarian insufficiency, and post–birth control syndrome.

LAB TESTING FOR LIGHT MENSES

Hormone tests can determine your levels of estrogen during your menstrual cycle. It may be helpful to collect samples of your hormones throughout the month with DUTCH Cycle Mapping or other hormone tests. The following are some routine tests you may also want to consider:

- estradiol, FSH, and LH: On day 3 of your menstrual cycle, have your estradiol tested.
- progesterone: On days 19–22 of your cycle, have your progesterone tested.

WHAT TO DO RIGHT NOW

If you're eating a low-fat diet—stop! Fat is your friend and how you build your hormones. Use the guide to healthy fats on page 223 to begin incorporating hormone-supportive fats into your diet.

If you've been on the pill for any period of time, consider a prenatal or multivitamin (see chapter 12 for recommendations). Key nutrients needed for making estrogen and balancing hormones are depleted by the pill.

Late or Irregular Periods

Never know when your period is coming? Does it seem to just show up whenever it pleases? There can be several reasons for a delay in menses or irregularity with your period, but the number one reason we always want to rule out is pregnancy. If you're sexually active and your period is late, please take a quick home pregnancy urine test to be sure—and don't be misled into thinking that just because you're using hormonal birth control that you can't get pregnant, because it can and does happen. If you want to know how effective your birth control method is, see the chart in chapter 13 (page 284).

Stress, in the form of physical or mental trauma, can cause a delay in ovulation and your period. When stress is very high, a woman may miss her period altogether. Other conditions that can contribute to this include PCOS, perimenopause, postpartum, diabetes, and celiac disease.

LAB TESTING FOR LATE OR IRREGULAR PERIODS

Testing the following can help determine the cause of your delayed or irregular period:

- adrenal testing (4-point salivary cortisol or 4-point urinary cortisol and cortisone with DHEA-S)

- DHEA-S

- estradiol, FSH, and LH: On day 3 of your menstrual cycle, have your estradiol tested.

- fasting insulin, fasting blood glucose, and hemoglobin A1C

- *pregnancy test*

- progesterone (taken about 5 to 7 days after ovulation; you'll need to use an ovulation-prediction kit to time this)

- SHBG

- testosterone (total and free)

- thyroid panel: TSH, total T4, total T3, free T4, free T3, reverse T3, anti-TPO, and anti-thyroglobulin antibodies

- vitamin D

WHAT TO DO RIGHT NOW

Using the dietary and lifestyle practices of the 30-Day Brighten Program, along with the right supplements, can help you begin to regulate your cycle.

Short Cycles

If you see your period sooner than every twenty-one days, then it's time to ask your body for a little more space. The soonest you should see your next period is twenty-one days, but if we're talking a healthy cycle, that will be at least twenty-six days. That's from day one of your first period to day one of the next. Shorter than twenty-one days is what is known as a luteal phase defect. You may not have ovulated or it may be that your corpus luteum didn't form correctly, which leads to a deficiency in progesterone. If you haven't already, take the Hormone Quiz in chapter 1 (page 16) to determine if you have symptoms of low progesterone.

LAB TESTING FOR SHORT CYCLES

If you are experiencing short cycles, consider having the following tested:

- DUTCH test
- estradiol, FSH, and LH: On day 3 of your menstrual cycle, have your estradiol tested.
- ovulation predictor kit (OPK): On days 10–15 of your cycle, use an OPK to determine if you are ovulating.
- progesterone: Test your progesterone five to seven days after ovulation.
- thyroid panel: TSH, total T4, total T3, free T4, free T3, reverse T3, anti-TPO, and anti-thyroglobulin antibodies
- 21-hydroxylase antibodies
- vitamin D

Tracking basal body temperature daily can provide insight into potential ovulation. See chapter 13 on the fertility awareness method and how to track this important data. Your doctor may recommend an ultrasound to see if ovulation occurred and if a corpus luteum is present. He or she may also recommend an endometrial biopsy to evaluate the effect progesterone has on the uterine lining.

WHAT TO DO RIGHT NOW

If you aren't already tracking your cycle, then start now. You'll want to get your symptoms, including ovulation, dialed in so you can understand what may be happening.

Using the herbal supplement Vitex, or chaste tree berry, starting the day after ovulation until two days after menses begins can help some women lengthen their cycle. Vitex works by stimulating the communication between the brain and the ovaries to encourage more progesterone production. You'll need to use this for at least three months to see an effect. Some women see changes after one cycle, but most women really take note after three to four months.

The health of your adrenals may also be the cause of low progesterone. Consider starting Dr. Brighten Adrenal Support in order to nourish your adrenals and support healthy progesterone. You can learn more about adrenal health in chapter 7.

A Missing Period (Amenorrhea)

Amenorrhea is classified as either primary or secondary: primary amenorrhea is when you have never had a period and are fifteen years or older; secondary amenorrhea is when your period has gone missing for more than three months if you've had regular menstrual cycles or six months if you've had irregular menstrual cycles. The most common causes of secondary amenorrhea include the following:

- **Post–birth control syndrome (post-pill amenorrhea):** We covered this in detail in chapter 3, and you may recall that some women lose their period after stopping the pill.

- **Functional hypothalamic amenorrhea:** Twenty-five to 35 percent of secondary amenorrhea cases are due to a reduction in gonadotropin-releasing hormone (GnRH). GnRH is what tells the pituitary to make FSH and LH, which result in egg maturation and ovulation. But when GnRH is misfiring, women don't experience development of their follicle, mid-cycle LH surges (remember, this is what triggers ovulation), and estrogen levels are typically low.

- **Pituitary dysfunction:** This occurs in roughly 17 percent of cases, with the most common issue being hyperprolactinemia, or too much prolactin hormone. Prolactin is the hormone that causes women to produce breast milk, but it can also escalate with hypothyroidism or with a prolactinoma, a benign brain tumor.

- **Ovarian dysfunction:** This occurs in about 40 percent of women who lose their period, with 30 percent due to PCOS and the remaining 10 percent due to primary ovarian insufficiency. This also can be autoimmune in origin, so one place to start is with 21-hydroxylase antibody testing, which is how Addison's disease is diagnosed too (Addison's disease can also attack the ovaries). Your doctor may assess your ovarian reserve using anti-Müllerian hormone (AMH) and follicular count via a transvaginal ultrasound. These two markers provide a lot of insight into a woman's fertility and aging of the ovaries. Women nearing menopause have low values on these tests.

- **Hypothyroidism:** While it's estimated that this is responsible for only 1 percent of amenorrhea cases, it's important to recognize that hypothyroidism is underdiagnosed in women. If you've lost your period, then having complete thyroid testing is a must.

LAB TESTING FOR A MISSING PERIOD

If you've lost your period, I recommend considering the following tests:

- brain MRI
- celiac panel: tissue transglutaminase antibody (tTG-IgA), endomysial antibody (EMA-IgA), deamidated gliadin peptide antibodies (DGP–IgA and IgG), and total serum IgA
- Cyrex Array 5 antibody testing
- DUTCH adrenal or DiagnosTechs ASI
- free and total testosterone and SHBG
- hemoglobin A1C, fasting insulin, and fasting blood glucose
- random FSH, LH, and estradiol
- *pregnancy test*
- prolactin
- thyroid panel: TSH, total T4 and T3, free T4 and T3, reverse T3, anti-TPO, anti-thyroglobulin antibodies
- 21-hydroxylase

WHAT TO DO RIGHT NOW

While you're working on getting your labs, begin the 30-Day Brighten Program, along with taking a multivitamin, to ensure you're getting the nutrients your body needs to build hormones and repair tissues. I also recommend seed cycling (see page 258) and fine-tuning your circadian rhythm (see page 233).

Mid-Cycle or Premenstrual Spotting

While mid-cycle spotting may be benign, it also may be a sign of something much more serious, like fibroids, infection, endometriosis, or cancer. It can be a sign of pregnancy too, so consider taking a test if you are experiencing mid-cycle spotting and are concerned:

- pregnancy test
- progesterone: On days 19–22 of your cycle, have your progesterone tested.
- thyroid panel
- transvaginal ultrasound

It's important to have this investigated by a doctor. While you're working on your root cause, try taking the herb Vitex and beginning Adrenal Support to optimize cortisol and progesterone production. Also focus on progesterone support, found on page 240.

Pain or Bleeding with Sex

No, pain with sex is *not* normal. Yes, you are meant to enjoy sex. Unfortunately, many women who are on the birth control pill develop vaginal dryness and other issues that can make sex, well, less than enjoyable. Dyspareunia—or pain with sex—can occur for several reasons. If you experience discomfort or burning with sex, it may indicate low estrogen, a yeast infection, or another type of infection. Vaginismus is a spasming of the vagina's muscles that can make intercourse painful. Generally, women with this condition also have pain when inserting a tampon and during gynecological exams. Vulvodynia is a chronic pain condition affecting the outer genitalia. Research has shown that women who begin the pill before age sixteen are nine times more likely to develop vulvodynia compared to women who have never taken it. This is due to the alteration of natural hormones. If you have pain with deep intercourse, this could be an issue of anatomy and position, but it may also be a sign of endometriosis, ovarian cysts, fibroids, infection, or another condition. If you are experie nificant bleeding after intercourse, this may be a sign of infection, ca inflamed cervix. I recommend scheduling a visit with your doctor to u're using lubrication and plenty of it, and that it's a green—or clean—kind. If you're having vaginal dryness, try a 400 IU vitamin E suppository. Poke a hole in the top of a suppository and insert the vitamin E into the vagina to help those cells plump up and be healthier. *Note: This will cause a condom to fail.* I also recommend making use of the information contained in this book once you know your cause. Meeting with a pelvic floor physical therapist can be highly beneficial too.

PMS

PMS—the term that has been at the center of so many jokes, yet is no joke if you're dealing with it. PMS symptoms include cramps, bloating, breast tenderness, headaches, trouble sleeping, sugar cravings, irritability, anxiety, depression, and mood swings. It's a syndrome, which means that you can have some or all of these symptoms, just like PBCS. And just like PBCS, there are many critics who still suggest PMS isn't real.

PMS can be debilitating, and while *you* get to blame it on your hormones, no one else better say a thing about your hormones when you're PMS-ing. I, like many women, used to think PMS is a normal part of being a woman. But it doesn't have to be. The problem with ignoring these symptoms and popping an ibuprofen or taking the birth control pill for them is that you don't get to the root of why you have them. PMS stems from a hormonal imbalance, usually estrogen dominance, which we'll explore next.

WHAT TO DO RIGHT NOW

Estrogen will cause you to hold on to water weight. If you find yourself retaining water or feeling bloated or kind of puffy before your period, dandelion leaf can help you naturally expel the excess water. Dandelion leaf can be drunk as a tea, eaten in a salad, or used as a tincture. As you'll learn in chapter 7, your adrenal health is the foundation of sex hormone health. Adrenal Support can help optimize adrenal function to balance cortisol and progesterone production.

If you're feeling anxious, couple some deep breathing with a passionflower tincture. Ingest 1 to 2 dropperfuls of the tincture when you feel that anxiety start to crop up. If a low mood is your problem, get out with your friends and move your body so you stay motivated and use some endorphins to combat what's going on with your hormones.

If you're having trouble sleeping before your period, try one of my supplements, Adrenal Calm, which contains phosphatidylserine, which has been shown to lower cortisol, and herbs that nourish the parasympathetic ("rest and digest") nervous system. If you're craving sugar, salt, or carbohydrates like a fiend the week or two before your period, head over to chapter 7 and show your adrenals some love stat. It's also a sign that you should be consuming more protein, which will provide you with tryptophan, a precursor to serotonin, a sleep neurotransmitter. Getting adequate amounts of B vitamins will also help. And if those sugar cravings are super intense and getting the best of you, then see page 226 for sugar craving strategies.

If you suffer from monthly PMS symptoms, a number of natural treatments can help. A diet high in sugar, refined carbs, non-organic meats, dairy, and caffeine can exacerbate your PMS, but a diet low in these foods combined with more vegetables, healthy fats, and fiber will assist in balancing your hormones and reducing your symptoms. Do the following to help eliminate your PMS:

- **Take magnesium glycinate.** I generally recommend 300 milligrams twice daily seven days before menses and for the first three days during. The rest of the month, aim for 300 milligrams nightly.

- **Add fiber.** Eat at least 25 grams of fiber every day.

- **Lower your inflammation.** Inflammation increases the activity of aromatase, an enzyme that converts testosterone to estrogen. With higher estrogen and lower testosterone, you will feel cranky, weepy, and unmotivated—the mood-related symptoms of PMS. (Side note: your libido will also disappear.) How do you lower inflammation? By consuming an anti-inflammatory diet, turmeric, and omega-3s (more on this in chapter 6). Besides improving your mood, lowering your inflammation will also help with cramps, lower back pain, and fatigue.

- **Handle your stress.** Stress can have a huge impact on the body—and on gut health, as you'll see in chapter 6. See the stress-reducing practices detailed on page 260.

- **Regulate your insulin.** Cut out foods that are high in sugar and simple carbs, and instead eat healthy fats and protein. Dietary fiber also helps regulate your insulin and blood sugar, as will regular strength training.

- **Try Vitex.** Take Vitex from day 15 to 28 of your cycle.

- **End estrogen dominance.** See the next section.

What Is Estrogen Dominance?

If you're suffering from serious PMS symptoms, there's a pretty good chance you have estrogen dominance. Estrogen dominance is one of the most ubiquitous hormonal imbalances I see in my patients, because both stress and environment have a major impact on your hormone levels. Women often think it's something they won't have to worry about before perimenopause. But estrogen dominance is really common and can affect any of us—at any time.

Estrogen in and of itself is not *bad*. No single hormone is bad; they all just need to be in balance. When you have too much estrogen, you may experience headaches, weight gain, tender breasts, heavy periods, irritability, and mood swings. Estrogen dominance is the underlying cause of why you gain weight in your hips, butt, and thighs, because estrogen stimulates your fat cells to store more fat, plus it leads to decreased amounts of available thyroid hormone.

There are two types of estrogen dominance: frank and relative. Frank estrogen dominance is when you simply have too much estrogen, period. Relative estrogen dominance is when you have too much estrogen relative to the amount of progesterone, which may be due to anovulation or adrenal issues.

WHAT CAUSES ESTROGEN DOMINANCE?

We live in an interesting time. We're being hit with more chemicals than the human body has ever experienced from a generational standpoint. That hot red lipstick you put on this morning? Chemicals. That shower gel you love so much that smells like vanilla? Chemicals. I know, it's a real bummer. Between our skin cleansers, cosmetics, and hair products, we women are exposed to so many

---

The Pill and Weight Gain

When you started taking the pill, did you also start gaining weight? I know I sure did! Maybe at first it was gradual and then you started having trouble buttoning your jeans. Then one day you stepped on the scale and realized, *Holy crap, I've gained 10 pounds!* Research says that the birth control pill causes minimal weight gain, but many women report having this side effect, and it's actually listed as a side effect in the package inserts. The weight gain is often due to fluid retention . . . and all that estrogen.

Alcohol and Estrogen Dominance

Please don't shoot the messenger on this one, but alcohol does affect estrogen dominance. In fact, one alcoholic drink can increase estrogen by 10 percent. It bogs down your liver, and your liver is responsible for processing all that estrogen, which it will have to put on hold while it processes the alcohol first. When your hormones are off, sometimes it feels like the only thing that saves your day is that glass of wine in the evening, which can make your estrogen levels worse. If you can get your hormones in balance, you won't need that alcohol for stress relief. I promise.

chemicals every single day. We're exposed to them in our personal care products, in the environment, and even in our food supply.

These chemicals are one of the primary causes of estrogen dominance. Diet is extremely important, because the pesticides sprayed on fruits and vegetables hate your hormones. They beat them up and block receptors and are big old bullies. Meat (and soy) sources can also contribute to estrogen dominance, because farmed animals are given lots of inflammatory grains, hormones, and antibiotics, which get passed on to you when you eat them. You truly are what you eat: what you consume becomes every cell of your body. Try to get your produce from local organic farms or grow your own, and always buy grass-fed organic meats to avoid exposing yourself to hormone-disrupting chemicals. Also, avoid buying food stored in BPA-laden cans, drinking out of plastic water bottles, cooking with Teflon, and using plastic spatulas, which can all introduce more chemicals into your body. Most of the things that probably make cooking easier come with a price (see Resources for hormone-friendly cooking alternatives).

The other causes of estrogen dominance, including stress and gut health, can be addressed by taking the following steps.

STEPS TO BALANCE YOUR ESTROGEN RIGHT NOW

Although estrogen dominance can make you feel miserable, it is reversible. Beyond paying attention to what chemicals you're exposed to in your environment, I recommend doing the following. First and foremost, reduce stress and focus on getting enough sleep (steps 4 and 5). By implementing all these steps, you will improve your estrogen-progesterone balance and begin to heal your entire hormonal system.

1. **Fix your gut.** Once the liver processes estrogen for elimination, it's up to the gut to move it out. If you're constipated, there's a good chance that isn't happening and your estrogen is going back into circulation in your body. (Again, I'll cover how to heal your gut in chapter 6.)

2. **Love your liver.** Your liver is responsible for preparing estrogen to be moved out of the body. Be sure to eat quality protein, plenty of garlic and onions, and a minimum of 3 cups of cruciferous vegetables weekly. Also, take a B complex vitamin. (For more on how to love up your liver, see chapter 5.)

3. **Eat fiber.** Aim for at least 25 grams of fiber per day by eating plenty of vegetables and fruits. Fiber will keep your bowels regular and help eliminate waste, including unnecessary estrogen. (Chapter 6 details fiber's benefits.)

4. **Stress less.** I know, easier said than done! But all that stress is wrecking your mood, your hormones, and your life. When you're stressed, you pump out more cortisol, which means you don't make as much progesterone, and you now have relative estrogen dominance. When you're super stressed out, you might not even ovulate—or you might ovulate later in your cycle. This is how powerful stress can be on your body. (See chapter 9 and "Stress-Reducing Practices" on page 260 for how to reduce stress.)

5. **Sleep.** Make sure you get at least seven hours of sleep every night. Your body needs it and your hormones demand it. You may want to consider wearing blue-light-blocking glasses one to two hours before bedtime to help you rest and shift your hormones to a more sleep-favorable state. (See chapter 11 for resetting your circadian rhythm.)

6. **Take an estrogen-supporting supplement.** Consider a combination product that provides you with calcium-D-glucarate, DIM (diindolyl-methane), and broccoli seed extract, which will help move your estrogen onto the right pathways out of your body. (Try my own version, Balance; see Resources, page 333.)

To figure out if your symptoms could be a sign of estrogen dominance, take the quiz on page 16.

Remedies for Period Problems

Period Problem	Supplement Type	How to Take	Recommended Brands
Estrogen dominance	DIM	100 mg daily	Dr. Brighten Balance or Integrative Therapeutics
	Calcium-D-glucarate	400 mg daily	Dr. Brighten Balance or Xymogen
	Probiotic	As directed	Microbiome Labs, Klaire Labs, or Designs for Health
Heavy or long period	Iron bisglycinate	18–30 mg daily	Dr. Brighten Prenatal Plus, Designs for Health, or Thorne
	B-vitamin complex	As directed	Dr. Brighten B-Active Plus, Innate, or Designs for Health
	Turmeric	1 g daily	Dr. Brighten Turmeric Boost or Integrative Therapeutics
	DIM	100 mg daily	Dr. Brighten Balance or Integrative Therapeutics
	Calcium-D-glucarate	400 mg daily	Dr. Brighten Balance or Xymogen
Painful period	Magnesium (glycinate or citrate)	300–600 mg daily	Dr. Brighten Magnesium Plus or Klaire Labs
	Cramp bark	1 teaspoon two or three times daily two days before your period and for the duration of your cramps	Wise Woman Herbals
	Omega-3	1,500–2,000 mg daily	Dr. Brighten Omega Plus or Nordic Naturals ProEPA Xtra
	Turmeric	1 g daily	Dr. Brighten Turmeric Boost or Integrative Therapeutics

Remedies for Period Problems

Period Problem	Supplement Type	How to Take	Recommended Brands
Light period	Prenatal or multivitamin	2 capsules twice daily	Dr. Brighten, Innate, or Seeking Health
	Vitamin D	2,000 IU daily or per lab findings	Dr. Brighten Vitamin D3/K2 or Thorne
Late or irregular period	Comprehensive hormone support (B6, B12, folate, DIM, broccoli extract, calcium-D-glucarate, green tea extract, black cohosh, Vitex, resveratrol, magnesium, and chrysin)	2 capsules twice daily or as directed	Dr. Brighten Balance, Thorne, or Integrative Therapeutics
Short cycle	Vitex/chaste tree berry	As directed in a combination product or as a tincture 60 drops twice daily in the luteal phase	Dr. Brighten Balance or Wise Woman Herbs
	Adrenal support	3 capsules in the morning	Dr. Brighten Adrenal Support
Missing period	Multivitamin or prenatal	As directed	Dr. Brighten, Innate, or Seeking Health
	Start the Brighten Supplement Protocol (see page 246)		

Remedies for Period Problems

Period Problem	Supplement Type	How to Take	Recommended Brands
Mid-cycle spotting	Vitex/chaste tree berry	As directed in a combination product or as a tincture 60 drops twice daily in the luteal phase	Dr. Brighten Balance or Wise Woman Herbals
	Adrenal support	3 capsules in the morning	Dr. Brighten Adrenal Support
PMS	Dandelion leaf tincture (water retention)	60 drops twice daily	Wise Woman Herbals
	Passionflower (anxiety)	60 drops of tincture	Wise Woman Herbals
	Adrenal Calm (sleep disturbance)	3 capsules at night	Dr. Brighten Adrenal Calm
	Magnesium glycinate	300–600 mg at bedtime	Dr. Brighten Magnesium Plus or Klaire Labs
	B-vitamin complex	1 capsule daily	Dr. Brighten B-Active Plus, Innate, or Designs for Health
	Comprehensive hormone support (B6, B12, folate, DIM, broccoli extract, calcium-D-glucarate, green tea extract, black cohosh, Vitex, resveratrol, magnesium, and chrysin)	As directed	Dr. Brighten Balance, Thorne, or Integrative Therapeutics
	Adrenal support	3 capsules in the morning	Dr. Brighten Adrenal Support

Key Takeaways:
Take Back Your Period

- The most common reasons for heavy or long periods are estrogen dominance and thyroid disease.

- If you have heavy and/or long periods, incorporate iron-rich foods into your diet as well as foods rich in B12, B6, and folate.

- Significantly painful periods can be a sign of infection, endometriosis, fibroids, or ovarian cysts and shouldn't be ignored.

- Magnesium, cramp bark, fish oil, vitamin E, and thiamine can all help with period cramps.

- If you have short or light periods, it may stem from a low-fat or vegetarian diet or overexercising.

- Stress can cause late or irregular periods—but so can pregnancy!

- Vitex, or chaste tree berry, can help lengthen a short cycle and reduce symptoms of PMS.

- The most common reasons for a missing period include post–birth control syndrome, functional hypothalamic amenorrhea, pituitary dysfunction, ovarian dysfunction, and hypothyroidism.

- Mid-cycle spotting can be a sign of fibroids, infection, endometriosis, cancer, or pregnancy.

- Estrogen dominance is really common, and what you're exposed to in your environment is one of the primary causes, along with stress and gut health.

- The six steps to balancing your estrogen are: fix your gut, love your liver, eat fiber, stress less, sleep, and take an estrogen-supporting supplement.

PART II

YOUR BODY ON THE PILL

CHAPTER 5

BIRTH CONTROL HORMONE DETOX 101

My patient Maya came to me when she was twenty-seven years old. She had started the pill when she left for college because her doctor had said it was the responsible thing to do. After going to college and not becoming sexually active, she thought, *Why am I on this?*, and she decided to stop taking it. Three months later, her skin was uncontrollably oily, and six months later, she had the worst case of cystic acne and "backne." Maya was on the swim team, so she was super embarrassed that her entire back was covered in acne. In desperation, she finally went to a dermatologist and started taking Accutane. She also wound up taking antibiotics for a period of time but found that nothing was really clearing her acne for good. She began the pill again in a moment of desperation when her gynecologist suggested that getting back on the pill would fix her skin. What happened? Her skin didn't improve much at all. The acne persisted, and Maya's mood began to tank. She was no longer motivated to do the things she loved.

I started Maya on the 30-Day Brighten Program, which includes a two-week liver detox. She was thrilled that her skin improved 80 percent, and she was ready to transition off the pill. After she finished her pill pack, she repeated the liver detox, which finally cleared the "backne" she had struggled with for years. The pill had disrupted Maya's hormones and burdened her liver, and when she came off it the first time, she experienced an androgen rebound (which we'll discuss further in chapter 8), all of which contributed to her acne.

When my patients come to me with hormonal imbalances or symptoms of PBCS, like oily skin and acne, digestive upset, raging PMS, or heavy, painful periods, the liver is one of the first areas we address. It's key to metabolism, detoxification, nutrient absorption, blood sugar balance, and immune system function. The liver is kind of the unsung hero of the body. We all think of it for detoxing alcohol and medications, but we forget that it also processes our hormones.

The liver is one of the organs of detoxification that eliminates hormones the body no longer needs—and that includes synthetic hormones from the birth control pill. This can mean the liver gets pretty "stressed out" while you're on the pill. Supporting the liver with a detox protocol is an essential step in the recovery of your hormonal health, and it's a quick way to start feeling better fast. I recommend doing a liver detox right away, as this can move the needle the most in terms of hormonal health. If you can do nothing else, hitting a detox for two weeks will improve your mood and energy and help with weight loss. And if you're staying on the pill, a detox every three to four months is in order.

Detox isn't just about what the liver does and how the body moves things out; it's also about what comes in. You have to look at the whole cycle: what you're exposing yourself to, how your body is processing it, and how your body is eliminating it. Besides some of the usual culprits already mentioned, the liver today has a much greater burden than it did a hundred years ago. From the foods we eat to the beauty and cleaning products we use, the liver is bombarded with chemicals it has to process day in and day out. Because a true detox focuses on more than just the problematic foods, drinks, or drugs you're consuming, it helps to also reduce these environmental toxins, as I mentioned when discussing estrogen dominance. In this chapter, I'll provide you with a detox protocol for happy hormones that your liver will love.

Your Liver and Estrogen Metabolism

There are three main places where women get their estrogen. One is the ovaries, which are responsible for about 80 percent of our estradiol and about 10 percent of the lesser estrogens (estriol and estrone). The second source is the adrenal glands, by way of DHEA. And this is why adrenal health is so important when

In This Chapter

- The liver's role in hormone health

- How the pill can stress out your liver

- Why a 14-day liver detox can fast-track your recovery (on or off the pill)

- The best foods to support your liver

- Why you may want to toss your lipstick and lotion

- How your liver may be blocking you in the bedroom

we transition into menopause, because once the ovaries quit, it's up to the adrenal glands to kick out that DHEA so we still have some estrogen available. The third is your body fat, which is why a healthy body composition is so important to our hormone health.

Now, whether your estrogen is coming from your ovaries, adrenal glands, or fat cells, it's up to your liver to make sure the estrogen you don't need gets packaged up and prepared for elimination through the bowels. As you probably guessed, this also applies to any estrogen-based medication or other hormones. That's why gut function is crucial to hormonal health and recovering from the pill. Remember, you have to poop every day to get rid of your excess estrogen!

How the Liver Metabolizes Estrogen

In the liver, detoxification enzymes (CYP, to be exact) convert estrogen to 2OHE1, 4OHE1, and 16OHE1 metabolites by a process called hydroxylation. The most beneficial metabolite is 2OHE1, whereas 4OHE1 and 16OHE1 are associated with swollen, tender breasts, clots in your menstrual blood, and cancer. Yeah, the stuff none of us want to be dealing with. To get your estrogen into the more favorable 2OHE1 form you need to be eating cruciferous vegetables. Supplementing with DIM, resveratrol, quercetin, and liposomal glutathione can also help (see a more complete list in the "Complete a 14-Day Liver Detox" section, page 96).

After these metabolites are created, they must be processed and excreted through the gut or kidneys, otherwise they go back into circulation and push your PMS off a cliff. Your 16OHE1 can be turned into estriol to be excreted, but the other metabolites require a bit more work. In a process known as methylation, your COMT enzyme (catechol-O-methyltransferase reduces DNA damage) changes your 2OH and 4OH to water-soluble versions, which requires magnesium and B vitamins—yes, the very nutrients the pill depletes. These metabolites are then sent to your kidneys to be removed from your body. Your estrogen is also processed via conjugation in the liver, which enables it to be carried out in bile via your poop. If you're constipated or have the wrong critters growing (a bacterial imbalance can lead to elevated beta-glucuronidase, which reactivates estrogen), then you can find yourself dealing with more estrogen than your body intended. An overflow of estrogen has been associated with PMS, painful breasts, and heavy periods, and it plays a role in the risk of breast, cervical, and endometrial cancer. Eating fiber-rich foods and taking calcium-D-glucarate can help you keep your estrogen in check. And keep pooping. You gotta be pooping.

Because the pill depletes nutrients necessary for healthy estrogen metabolism, excess metabolites can build up and cause problems in your body. The individual differences in both genetics and environment is a big reason why some women do fine on the pill while others feel completely wrecked by it. It's also the reason why some of us start out fine on the pill and eventually develop major issues. The 30-Day Brighten Program is like a sheepdog that herds the estrogen into the right pathways using food, supplements, and lifestyle therapies so that you can handle your estrogen appropriately—on or off the pill.

Is the Pill Taxing Your Liver?

If you don't think you need to support your body's natural detox abilities, then you need to wake up to the reality of our environment, because today we all are exposed to more chemicals than ever before. They're hitting us left and right, though no one talks about the significant environmental toxin known as hormonal birth control (which is changing the gender of fish, so just imagine what it's doing to humans). You try to avoid eating out of plastic, to use more

Castor Oil for Your Liver?

A castor oil pack to support detox is old-school naturopathic medicine. Castor oil works as a counterirritant and increases blood flow and circulation. If the thought of a spoonful of castor oil makes you cringe, take a breath because this is topical only. A castor oil pack enhances liver function, stimulates the healthy flow of lymph fluid, and supports natural detoxification pathways. You can rub a small amount of castor oil directly over your abdomen in a clockwise direction and wear a white shirt to bed that you don't mind staining. You can also make a castor oil pack by folding a piece of flannel three layers thick, saturating it with castor oil, and placing the pack across your abdomen with an old towel and hot water bottle on top. You can use this pack repeatedly and simply add more oil to it when the flannel begins to feel dry.

environmentally friendly household products, and to make sure that you're doing all the other things right, and then you pop this pill every day. It's too much for your body. Your liver takes a hit while you're on the pill, and research is questioning if it ever recovers its original function.

Yeah, you read that right: the pill is messing with your liver's genes. Your liver produces sex hormone–binding globulin (SHBG), which binds excess hormones in your body. When you're on the pill, this protein increases to protect your body from the synthetic hormones. A study in the *Journal of Sexual Medicine* revealed that women who had been on the pill for at least six months had higher levels of SHBG than women who had never taken it, and these levels remained elevated several months later. Women who opted to remain on the pill had about four times the normal amount of SHBG! While these levels may eventually decrease, the researchers speculated that *they may never return to pre-pill levels.* There is some concern that long-term exposure to the synthetic estrogen in birth control pills actually alters your liver genes to make higher levels of SHBG for the rest of your life. Unfortunately, SHBG also binds up your testosterone. The result is a libido that's nonexistent, that's never to be seen again. And your libido and your orgasms are super important, which is why I prescribe orgasms all day every day in my clinic, and I'm going to be prescribing them for you as well and helping you get your libido back in chapter 10.

Top Liver-Loving Foods

Make sure you eat some of these top liver-loving foods every single day:

- beets
- broccoli and other cruciferous vegetables
- burdock or gobo root
- complete protein
- dandelion root tea
- garlic
- grapefruit
- green tea
- leafy green vegetables
- turmeric

Liver Tumors and Liver Cancer

It's no secret in medicine that benign liver tumors come with taking the birth control pill. Yes, I'm going to guess your doctor didn't tell you that, just like my doctor didn't tell me. And while much of the research says these tumors rarely become malignant, which leaves a lot of people to think, *Hey, there's nothing to worry about*, the reality is that these tumors have a high risk of bleeding or rupturing. And when women are on the pill we see multiple, larger tumors that are more prone to bleeding. When the body is sending about 27 percent of all its blood through the liver, any bleeding is big trouble!

The research on liver cancer is conflicting. Some research says that oral contraceptives place you at higher risk of developing liver cancer that's malignant, but other studies say there's no correlation and no increased risk. Right now we aren't totally sure what the risk is, but here's something I find really interesting: these benign liver tumors, known as hepatic adenomas, were rarely reported before birth control was introduced in the 1960s. Yes, let that sink in. More women are being diagnosed with benign liver tumors since the introduction of the pill.

There has been speculation that lower-dose, newer contraceptives come with a lower risk of developing liver tumors, but at this time there haven't been large enough studies to know for sure. Women over thirty using the birth control pill and long-term users are at increased risk of developing tumors. And when it comes to length of time, one study revealed that just six to twelve months of use resulted in about 10 percent of women developing tumors.

Most likely this is something you're now feeling concerned about, and the first step is to see your doctor. He or she can actually feel your liver. In some cases, it will be lumpy and bumpy, and that's not a good sign. Your doctor can also discuss your symptoms with you and order tests and imaging to investigate. I'll also teach you how to support your liver—on or off hormonal contraceptives.

Gallbladder Disease and the Pill

Your liver produces a quarter or more of the bile you need every day, which gets stored in your gallbladder. Remember, bile plus poop is one way you get your estrogen out. Estrogen dominance is a common reason why ladies lose their gallbladder, and the birth control pill can contribute to gallstones and gallbladder disease.

Women with gallbladder disease will feel nauseated or generally unwell after they eat things like fatty meals, pork, eggs, or onions, which tend to be big triggers. They can also experience indigestion, upper right abdominal pain, and even vomiting. Some women will just feel like their upper right shoulder blade is sore and aching—as if they have a backache—which is actually their gallbladder radiating pain there.

If you're wondering if you're at risk for gallbladder disease, here are the five f risk factors: fertile, female, forty, fair skin, and fat . . . or overweight (please remember this is a mnemonic, not a judgment). If you're a woman who is still fertile, you're already at risk, and then you throw the pill on top of it and ask your liver and gallbladder to deal with even more estrogen. Over time, your body becomes less effective at eliminating estrogen, plus your body begins to struggle to absorb fat-soluble nutrients, and your poop gets funky because of it. Now you've got some serious problems that can land you in surgery. But there are steps you can take to protect yourself! Drum roll . . . the 30-Day Brighten Program. See? I got you!

Environmental Toxins and Your Hormone Health

As I mentioned previously, most women are exposed to hundreds of hormone-disrupting chemicals each day before they even leave their homes. If you use common brands of shampoo, conditioner, hairspray, face wash, or body lotion with the frequency of an average American woman, you are exposing your body to hundreds of these chemicals daily, as well as chemicals that are known to be carcinogenic (cancer-causing agents). Yet there are few regulations when it comes to what manufacturers can add to these products. Many contain estrogenic substances—meaning they act like estrogen but are artificial hormones—which can wreak havoc on your body.

I often think back to how much beauty garbage I used to put on my body and it's pretty scary. I used vanilla-scented lotion every day in my twenties, sometimes multiple times per day. I don't know how I got it in my head that a woman should smell like vanilla, but I look back and cringe. I had no idea about the effects that these chemicals could have on me. As I became more educated about endocrine-disrupting chemicals, I removed all of these products from my life and replaced them with safer ones.

For a long time I was under the impression that what I put on my skin wasn't really getting into my body. But if you think about how hormones can be delivered topically—for example, some women use a topical progesterone—they are absorbed very well through the skin—your largest organ. The skin is great at protecting our insides from a number of things, but it can and will absorb chemicals.

I mentioned the lack of regulatory laws in personal care products. Many of the chemicals found in cosmetics have not undergone extensive testing, and unfortunately, the ones that have, have been tested usually on lab animals, not people. Many beauty products contain parabens, formaldehyde, and synthetic fragrances, which are all hormone-disrupting chemicals. Now think about this: an average of 90 percent of teenage girls begin applying makeup daily at age fourteen (about the time a girl gets her period). If she lives to the age of seventy, her body will have been exposed to these chemicals for more than fifty years. We simply do not know the long-term effects of exposing women of any age to these chemicals.

Besides the beauty and personal hygiene products you directly apply to your skin, the chemicals in the cleaning products in your home, whether it's the bleach you use to clean your bathroom, the liquid soap to wash your dishes, or the scrub for your tub, make their way through your skin and into your body.

As I mentioned previously, these chemicals are endocrine disruptors—hello, hormone imbalance!—and carcinogens, but they also have been shown to contribute to diabetes and obesity by directly affecting the metabolism of your thyroid hormone and binding to your receptors. It also has been theorized that environmental toxins shrink the thymus, the organ where immune cells go to mature and figure out what's you and what's not you. When the thymus shrinks, there's a diminished production of T regulatory cells, which have an important role in regulating the immune system and controlling autoimmunity. If the thymus shrinks and the T regulatory cells get diminished, there is a potential for the immune system to begin attacking tissues in the body, resulting in autoimmunity (more on this in chapter 6).

I realize at this point you're wondering how, exactly, you're supposed to clean your house and survive without your favorite products. Well, I've got you covered. See the following list of some alternative brands, but also, in the Brighten Detox Protocol, I'll arm you with everything you need to find safe household products and nontoxic sources for your beauty regimen. Bonus: if you follow the 30-Day Brighten Program in chapter 12, you'll start to have glowing skin from all the nutrient-dense foods you'll be putting into your body.

Safer Beauty Alternatives

Skin care: Annmarie Skin Care, The Spa Dr., Eminence

Moisturizer: coconut oil, calendula salve, FATCO

Makeup: bareMinerals, Jane Iredale, Vapour Organic Beauty

Deodorant: Schmidt's, Primal Pit Paste, PiperWai

Laundry: My Green Fills

For the most up-to-date personal products resources, please visit DrBrighten.com/Resources.

What Exactly Does It Mean to Detox?

The word "detox" tends to conjure up visions of juicing for days on end with no solid food or drinking nasty concoctions that resemble swamp water. But let's be clear: your body is detoxing every day through your liver, kidneys, gut, lymphatic system, lungs, and skin. Supporting detox is supporting the physiological process of metabolizing chemicals, hormones, and other environmental compounds and removing them from the body using nutrition, lifestyle, and targeted supplement therapy. This includes reducing incoming toxins, like alcohol, xenoestrogens, and other chemicals, that keep your liver working overtime. Now, it may sound like a lot of work to upgrade your diet and ditch some of these hormone-harming habits, but as you increase your energy, your mood, and your mojo, and create better hormone balance in your body, I'm pretty confident you'll say, "That was so worth it."

The Brighten Detox Protocol

If you're staying on the pill or if you've got period problems, you'll want to start this protocol ASAP. If you're coming off the pill, do this within the first month after you quit.

Your first step will be to eat the foods and nutrients your liver needs to process estrogen and other metabolic waste. The 30-Day Brighten Program contains a whole lot of liver-loving foods and eliminates those that can burden the liver or contribute to hormone imbalance. In my clinic, I also take women through a 14-day professional-grade supplement program as part of their detox (more on this soon).

You'll also replace any harmful chemical products you use on your body and in your home with safer, nontoxic versions, as well as avoid toxic people. You want to clean your body, mind, and home as much as possible of all toxins. Plus, you'll learn liver-supporting lifestyle practices that positively impact your hormonal health.

The key steps to the Brighten Detox Protocol are:

- Complete a 14-day liver detox.
- Eat liver-supporting foods.

- Eliminate liver-burdening foods.
- Reduce your toxic burden and create a toxin-free environment.
- Get yo body movin'.

Quiz: Does My Liver Need Detox Support?

You may be wondering how to determine if your liver could benefit from detox support. Check any of the following boxes that apply to you:

☐ I use prescription or over-the-counter meds (other than thyroid medication).

☐ I have a current or past history of taking hormonal contraceptives.

☐ I experience hormonal symptoms like PMS, breast tenderness, and general period problems.

☐ I experience fatigue and brain fog.

☐ I experience headaches and migraines.

☐ I get rashes, hives, acne, and/or itchy skin.

☐ I drink alcohol weekly or drink more than three drinks in one day.

☐ I eat non-organic meat and/or vegetables and fruits.

☐ I consume canned or farmed fish and seafood.

☐ I experience anger, aggression, or irritability.

☐ I get congested sinuses or postnasal drip.

☐ I have chemical sensitivities.

☐ I experience hypoglycemia or have a blood sugar imbalance.

If you checked two or more boxes, your liver needs some attention and you would benefit from added detox support. Go to chapter 12 to start the 30-Day Brighten Program, which includes a 14-day detox to reset your hormones.

Lab Tests for Liver Function

The unfortunate thing about liver function tests is that things have to be really bad for abnormal results to show up, because your liver is such a champ. Testing the following can help you better understand the health of your liver:

- alanine transaminase (ALT)
- alkaline phosphatase (ALP)
- aspartate aminotransferase (AST)
- bilirubin
- gamma-glutamyl transferase (GGT)

Complete a 14-Day Liver Detox

To balance your hormones and minimize side effects as you come off the pill, a 14-day liver detox is vital, which is why it is the starting place for the 30-Day Brighten Program. This will improve your energy, mood, and skin, and ease your periods. As I mentioned previously, if you stay on the pill, I recommend doing a detox every three to four months.

In my practice, we use either Dr. Brighten Paleo Detox or Dr. Brighten Plant-Based Detox, both of which include a variety of liver-boosting ingredients in easy-to-use packets. Here are the other liver-supporting supplements you can take on their own:

- **Liposomal glutathione** is the mother of all antioxidants. It's one of the fastest ways to love up your liver. Take 100 to 200 milligrams daily. If you're using glutathione on its own, I recommend the liposomal form because it actually binds to cells and facilitates the delivery of nutrients even more effectively.

- **N-acetylcysteine (NAC)** is a precursor to glutathione and does a whole lot of really great things like improving mood, fertility, and gut function, and may reduce the risk of miscarriage. Take 600 to 900 milligrams twice daily.

- **Milk thistle** has been shown to support the healthy regeneration of damaged liver cells and protect against liver damage. Take 300 milligrams three times daily.

- **Diindolylmethane (DIM)** helps your body process estrogen into safe metabolites and helps you maintain healthy levels of estrogen. Take 100 milligrams twice daily.

- **Calcium-D-glucarate** aids in liver detoxification and the elimination of excess estrogen. Take 50 to 1,000 milligrams daily, depending on the severity of your symptoms.

- **Dandelion root tea** specifically supports healthy liver detoxification and can make a great replacement beverage for people who want to kick coffee. Enjoy 1 to 3 cups daily.

- **Quercetin** is an antioxidant and anti-inflammatory found in red onions, blueberries, and chili peppers. It protects against oxidative damage and aids in phase I liver detoxification. Take 100 milligrams twice daily.

- **Resveratrol**, an antioxidant most associated with red wine (sorry, red wine won't cut it during your detox), aids in phase II liver detoxification, which may be protective against certain cancers. Take 100 milligrams twice daily.

Eat Liver-Supporting Foods

As I mentioned earlier, liver-loving foods contain the essential nutrients needed to support your natural detox pathways and are rich in DIM (diindolylmethane) and sulfur, which assist with the processing of estrogen and feed the good gut bugs (which help get the estrogen out). These foods are also full of minerals and vitamins that influence hormonal health.

Eating adequate fiber—a minimum of 25 grams daily—and drinking plenty of water will help you have regular bowel movements and remove excess estrogen from the body. Your detox diet should follow these guidelines:

- Aim for 3 to 6 cups of **organic vegetables** daily. Consume a variety to keep your meals interesting and support your body with a wide range of vitamins and minerals. Be sure to pile on those leafy greens, along with beets, carrots, garlic, onions, broccoli and other cruciferous vegetables, and artichokes.

- Your liver needs **high-quality protein** to operate its powerful detox pathways. Eat organic, 100 percent grass-fed meats and pasture-raised eggs as well as wild-caught fish, legumes, nuts, and seeds (these are especially important if you don't eat meat).

- **Hormone-healthy fats** help regulate your blood sugar and supply your body with the energy it needs to create hormones. You can find them in avocados, cold-pressed olive oil, coconut oil, macadamia nut oil, and olives.

- Proper **hydration** is an integral part of a detox. You should drink at least half your body weight in fluid ounces daily. Try to increase your intake by about 20 ounces per day during a detox.

Whenever possible, choose organic foods and 100 percent grass-fed meats. Avoid storing or purchasing food in plastic containers if possible. Placing plastic containers with fat-containing foods in the microwave is another source of xenoestrogens, which should be avoided.

Eliminate Liver-Burdening Foods

To assess how your hormones respond to certain foods, you will pull some of the common hormone-hating culprits. During your detox, do not consume the following foods or beverages:

- **Sugar:** There are many reasons to avoid sugar, including that it increases inflammation, disrupts your blood sugar level, and stresses your hormones, which is exactly what you don't want when you're trying to balance your hormones post-pill. And as I'll explain soon, you don't want anything increasing inflammation while you're on the pill.

- **Alcohol:** You may not be happy to hear this one, but remember, alcohol is a toxin. I'm sure you know that there is a correlation between alcohol consumption and liver health and that excessive amounts of alcohol can cause liver disease. During a detox you want to give your liver a break from processing alcohol and other toxins—it's only 14 days. (Although I recommend you take a full 30 as part of the 30-Day Brighten Program.) You can do it! Your liver—and hormones—will thank you.

- **Inflammatory fats:** Because you build your hormones from fats, if you want healthy hormones, you need to eat *healthy* fats—and avoid the fats that drive your hormones and immune system wild. These inflammatory fats include trans fats, canola oil, corn oil, cottonseed oil, peanut oil, and fats from conventional animal products.

- **Hormone-disrupting foods:** A number of foods disrupt your hormones. Give your body at least four weeks off from these foods and then slowly reintroduce them (we'll cover more of this in chapter 6). These foods include gluten and all grains, dairy, soy, and caffeine, as well as sugar, alcohol, and inflammatory fats. The only way to know how these foods affect you is to take them out and reintroduce them later.

Reduce Your Toxic Burden and Create a Toxin-Free Environment

Now that you know all the ways toxins can be harmful from earlier in the chapter, you can begin to reduce your body burden. Whether it's your lipstick or your lotion, you can replace these products with safer alternatives—and it's easier than ever before, with access to the Environmental Working Group's Skin Deep Cosmetics Database (visit EWG.org). Plus, if you missed it, check out my list of "Safer Beauty Alternatives" to try on page 93.

I also recommend doing what I call a mind–body detox. When you're detoxing, look at your relationships. Are they toxic? Look at the way you talk to yourself. Is that toxic? Look at the choice of words you use every day. Do you watch the news? Stop that! That's toxic. We are such complex biological and energetic systems that literally everything affects us. Focus on spending time with people who make you feel like glitter and less time with those who leave you feeling depleted, stressed, or [fill in the undesirable blank] when you're hanging out with them. Trust me, master this and you'll be unstoppable.

Get Yo Body Movin'

It makes sense that exercise helps you sweat out some toxins. If you've ever had a rough night of one too many cocktails but forced yourself to go to that early morning boot camp or hot yoga anyway, you know what I'm talking about. Exercise gets your lymphatic system moving and your blood flowing. And if you work up a sweat, it helps move waste out of your body. Aim for about thirty minutes of daily movement during a detox, with two to three days of sweat-inducing exercise like high-intensity interval training (HIIT), two days of strength training, and two to three days of yoga or another mind–body exercise.

Are you ready to kick off this detox? Then head on over to the 30-Day Brighten Program in chapter 12 and begin.

Key Takeaways:
Birth Control Hormone Detox 101

- The liver plays a crucial role in hormone balance and is essential in processing the pill.

- The pill may also permanently alter your liver genes to have higher levels of SHBG—say buh-bye, libido.

- The pill can cause liver tumors that are more prone to rupturing and bleeding.

- The birth control pill can contribute to gallstones and gallbladder disease. The five *f* risk factors are fertile, female, forty, fair skin, and fat (overweight).

- Our cosmetics and cleaning supplies can contain hormone-disrupting chemicals that *do* get absorbed by your skin and mess with your hormones.

- The Brighten Detox Protocol includes the following key steps: complete a 14-day liver detox, eat liver-supporting foods, eliminate liver-burdening foods, reduce your toxic burden and create a toxin-free environment, and get yo body movin'.

CHAPTER 6

GUT CHECK

Now, before you go trying to skip this chapter, thinking that your gut has nothing to do with your hormones, I want to tell you this: your gut has *everything* to do with your hormones, and it may be the very reason you're having to deal with symptoms like cramps, acne, headaches, irritability, anxiety, fatigue, or weight gain. I get it. I'm a doctor, and every day in my practice I hear, "Dr. Brighten, I just want to fix my hormones. Why are you talking to me about my gut?" In this chapter, I'm going to teach you just what the birth control pill does to your gut and why it needs some rehab if you're going to turn your hormones around.

So, there was this guy named Hippocrates who said "All disease begins in the gut" about 2,500 years ago, and you know what? He was right. Your gut is responsible for absorbing the nutrients from your food, which in turn help you make the hormones you need and break down and eliminate the ones you don't. It houses a whole lot of organisms—we're talking more organisms than there are cells in the body—that influence your mood, your weight, inflammation, and your overall hormonal health. In addition, your gut, as we've talked about previously, is responsible for moving hormones out of your body and helping you detoxify. Your gut is truly at the center of all the inner workings of your body, but when you're on the pill this system takes a hit, big time. From inflammation and leaky gut to a disruption of your entire microbiome, the pill makes a big impact and not in a good way. So if you're experiencing gas, bloating, constipation, or diarrhea, or you're seeing some things in your poop that don't look right, I'm going to help you understand why that is, how the pill is involved, and just what to do.

Your Microbiome, Gut, and Hormones Are All BFFs

Your gut houses your microbiome, which consists of healthy bacteria, or "good" gut bugs. When your gut isn't functioning properly, you can experience mood symptoms like anxiety and depression, headaches, joint pain, fatigue, difficulty with gaining or losing weight, and—you guessed it—hormone imbalance.

You need a healthy gut to remove excess estrogen and other waste from the body. Estrogen is moved out of the body through the bowels—this includes the estrogen you make naturally and the estrogen delivered synthetically through that daily dose of birth control. When gut function is compromised, estrogen goes back into circulation in your body, which can lead to those annoying symptoms like bloating, cramping, heavy periods, and irritability—or what is otherwise known as PMS. If your bowels aren't moving or you've got an imbalance in your microbiota (your good gut bugs), your estrogen sticks around longer than it should, contributing to a state of estrogen dominance. You have to poop every day to remove the estrogen your body no longer needs.

My patient Ava was suffering from severe PMS symptoms. She was experiencing extreme irritability and anxiety before her periods that would turn into full-blown panic attacks and screaming episodes followed by depression. She had trouble sleeping—her mind would race—the week before her period arrived. And the period itself? Heavy, like changing a super tampon every hour

Your Gut and Estrogen Dominance

Remember estrogen dominance from chapter 4? Well, guess what? Your gut can play a *big* role in estrogen dominance and those pesky PMS symptoms. Gut bacteria make an enzyme called beta-glucuronidase, which is beneficial to you . . . until it's not. Too much beta-glucuronidase means *more* estrogen. How? Because this enzyme undermines everything your liver did to prepare estrogen to move out of your body and instead causes that estrogen to go back into circulation. Now your body has way more estrogen than it needs. The result? Estrogen dominance. Getting your gut healthy can have huge benefits to your hormones. You can also take calcium-D-glucarate daily to help handle that beta-glucuronidase.

In This Chapter

- The pill, leaky gut, and your hormones
- Why you can't ditch those extra pounds
- The pill's role in autoimmune disease
- How good gut health can undo the effects of the pill and balance your hormones
- Lab tests for a personalized gut check

heavy. And painful. We're talking Midol popping, hot water bottle hugging, and out for at least a day, if not two, kind of pain. She also had gained weight steadily in her twenties, but she found that her weight really climbed after she started taking the pill, no matter how well she ate or how much she exercised. She just couldn't seem to lose weight in her butt, hips, or thighs.

Ava was also constipated—having bowel movements only every two to three days. She felt bloated most days, although she was passing very little gas. Ava's story provided clues that she was estrogen dominant, which is common with the pill. But since she was also constipated, the inability to remove estrogen through the bowels was further aggravating her already high estrogen state.

I ordered lab work to evaluate Ava, which included a stool culture, lactulose breath test, and blood work to evaluate the current state of her hormones, inflammation, and nutrients. Her labs revealed she was positive for methane dominant SIBO and bacterial and yeast dysbiosis, and had significant inflammation in the gut and systemically.

Ava wasn't ready to quit the pill. She didn't feel confident that she could effectively prevent pregnancy with the current options she had available. So we worked on reducing inflammation, improving gut function, and replenishing nutrient stores. Her symptoms began to improve, and her mood was not as extreme as it had been. Ava was able to lose some weight, and her digestion became more regular. But she couldn't entirely overcome her symptoms, which is when she ultimately decided it was time to ditch the pill.

If you have constipation like Ava, you're probably struggling with estrogen dominance. To help your bowels start moving right away, I recommend increas-

Quiz: How Is My Gut Health?

Check any boxes that apply to you in the following list:

☐ I have gas.

☐ I have bloating.

☐ I have discomfort and/or belching after meals.

☐ I have diarrhea.

☐ I have constipation.

☐ I have difficulty passing stools.

☐ I have food sensitivities.

☐ I have irritable bowel syndrome (IBS).

☐ I have Crohn's disease or ulcerative colitis.

☐ I have an autoimmune disease.

☐ I need stimulants (natural or drug) to have a bowel movement.

☐ I have heartburn.

☐ I take acid-blocking medication more than once a year.

☐ I experience nausea after meals.

☐ I am nauseous often.

☐ I have rashes, eczema, acne, or hives.

☐ I have foul-smelling stools.

☐ I see undigested food in my stools (other than nuts, seeds, or corn).

☐ I see mucus in my stools.

☐ I have cravings for sugar, simple carbohydrates, or alcohol.

☐ I experience anal itching.

☐ My tongue is swollen or has a thick coat.

☐ I use NSAIDs like ibuprofen or naproxen regularly.

☐ I have taken antibiotics more than once in the last year.

☐ I have taken the birth control pill.

TOTAL _____

Quiz: How Is My Gut Health?

Answer Key

1 or 2 boxes checked = high gut health (mild dysfunction or imbalance)

3 or 4 boxes checked = intermediate gut health (moderate dysfunction or imbalance)

5 or more boxes checked = low gut health (severe dysfunction or imbalance), and you would benefit from taking the supplements recommended in the "Nourish" section on page 116.

Whatever your gut health, follow my Brighten Gut Repair Protocol to improve your gut—it's the foundation to healing PBCS, rebalancing your hormones, and protecting your health while you're on the pill. In this chapter we are going to explore just how one little pill can wreck your gut.

---

ing your fiber intake, aiming to get at least 25 grams a day, taking a probiotic, and drinking plenty of water. A daily ginger supplement of about 1,000 to 2,000 milligrams or magnesium citrate at about 300 milligrams can help support regular bowel movements. This is all part of the 30-Day Brighten Program, which I prescribed for Ava. And if you have underlying gut infections, you have to address the root cause of what's going on as well.

How the Pill Wrecks Your Gut

Got a gut like Ava's? Beyond bowel movements, your gut and the critters it houses support your hormones and overall health. In fact, having more diverse gut flora (gut bugs) creates more favorable estrogen metabolites and may lower the risk of breast cancer. (The pill increases the risk of developing breast cancer—more on this in chapter 8.)

One of the major offenders that alters your microbiome is chronic medication use. Birth control pills disrupt normal flora (what should grow in your gut) and produce an environment that allows for overgrowth of harmful bacteria and yeast. But the signs and symptoms may be subtle—mild gas and belching, increasing food sensitivities—or they may be overt, like foul-smelling gas, bloating, constipation, and irritable bowel syndrome, and even symptoms of depression, acne, or eczema can indicate gut dysbiosis or microbial imbalance. I've often found in my practice that these symptoms will not completely resolve until the pill is discontinued. But the pill is doing more than messing with your microbiome; it's also destroying your *gut integrity*.

Research has shown that the pill can inflame the digestive tract and cause so much immune dysregulation that it increases the risk for autoimmune disease of the gut. During times of inflammation, intestinal hyperpermeability, or leaky gut, can develop. This is why healing your gut is an important step in balancing your hormones and recovering from PBCS.

What Is Leaky Gut?

Intestinal hyperpermeability, also known as leaky gut syndrome (LGS), is a condition that leads to widespread inflammation and hormonal disruption throughout the body. Yeah, it's as bad as it sounds. Leaky gut occurs when the tight junctions between the cells of the intestinal lining are compromised. These tight junctions regulate what is allowed in and out of the intestinal wall. The gut is naturally permeable to very small molecules so that the body can absorb the nutrients it needs. That's the way it's supposed to work. But in times of stress, and when you're taking the pill, if you're eating foods that you're sensitive to, you have a gut infection, or you take NSAIDs or antibiotics, then larger proteins and molecules from food, bacteria, yeast, and other organisms get through the intestinal lining that shouldn't be allowed in. Then toxins, microbes, and undigested food particles make their way through your intestinal lining, which sets off your immune system. When larger food molecules make their way past the gut barrier, the immune system recognizes them as "non-self" and goes on the defensive. The result? An inflammatory attack by your immune system that can lead to an array of uncomfortable symptoms. Leaky gut syndrome is how food sensitivities and immune disorders like autoimmune disease develop.

PCOS: What's Leaky Gut Got to Do with It?

There is a strong connection between the gut and hormone health, but new evidence suggests that the gut can also play a role in polycystic ovary syndrome

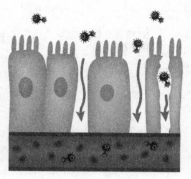

Normal Tight Junction   **Leaky and Inflamed**

A healthy gut is selective, allowing only necessary particles through and keeping pathogenic organisms and large food proteins out. When the tight junctions between the cells are compromised, undigested food, harmful bacteria, yeast, and parasites make their way in, which triggers the immune system and leads to inflammation and food sensitivities.

(PCOS). It is well understood that PCOS is rooted in inflammation and insulin dysregulation (more on this in chapter 8), but what actually triggers a woman to develop PCOS is less understood. New research is exploring the dysbiosis of gut microbiota (DOGMA) theory, which proposes that imbalanced gut flora and leaky gut trigger the immune system to create inflammation, which leads to insulin receptor dysfunction and higher levels of testosterone. This theory may explain why women with PCOS experience anovulatory cycles, irregular periods, acne, hirsutism, and ovarian cysts—and it all comes down to gut health. And it raises the question of whether giving the pill to "regulate cycles" may be doing more harm than good in women with PCOS.

The Pill: The Gasoline to Your Autoimmune Fire

The impact the pill has on autoimmune disease of the gut is profound—**there is a 300 percent increased risk of developing Crohn's disease if you take the pill!** The majority of your immune system is found in your gut, so if you've got a family history of autoimmune disease and you're taking a drug every day that has the potential to piss off your immune system, well, you're playing with fire. If you have a genetic predisposition to autoimmune diseases, it's possible they may be triggered. And if you're a woman who smokes and takes the pill, you have a significantly higher risk of developing ulcerative colitis.

What Causes Leaky Gut Syndrome?

- acid-reducing drugs
- antibiotics
- birth control pill
- BPA (bisphenol A)
- chronic stress
- dental and sinus infections
- dysbiosis
- food sensitivities
- gluten
- gut infections
- head injury
- heavy metals
- high-sugar diet
- HPA (hypothalamic-pituitary-adrenal) axis dysregulation, or adrenal fatigue
- low-fiber diet
- NSAIDs (non-steroidal anti-inflammatory drugs)
- nutrient depletion
- pesticides
- SIBO (small intestinal bacterial overgrowth)
- steroids
- surgery
- trauma

Evidence has shown that the pill is also associated with increased risk of lupus, an autoimmune disease that affects the joints, skin, and kidneys, and interstitial cystitis, commonly called painful bladder syndrome. A substantial review of hundreds of peer-reviewed studies related to hormonal contraceptives and autoimmune disease found that the combined (estrogen and progestin) pill was associated with an increase in multiple sclerosis too, as well as Crohn's disease, ulcerative colitis, systemic lupus erythematosus, and interstitial cystitis. Interestingly, there was a decreased risk in developing hyperthyroidism. Progestin-only pills were linked to an increase in eczema, hair loss, hives, and joint pain. To understand how the pill is involved in the development of autoimmune disease, let me catch you up on exactly what an autoimmune disease is and how it develops.

Mom's Microbiome Is Baby's Microbiome

What grows in a mom's gut grows in her baby's gut, so if you've got some critters that don't belong, or the pill left you with imbalances, this can impact your baby's gut in a big way. That's why I encourage women who are currently on the pill to take probiotics. And if you're thinking about becoming pregnant, do some of the work I recommend in this chapter to get your gut health in check so your body is baby ready.

The microbiome of a mother is so important that her baby's immunological health can be affected for life. For example, babies who are born via C-section and do not get exposed to their mother's vaginal flora by passing through the vaginal canal have higher incidences of celiac disease, diabetes, asthma, and other chronic conditions. Fortunately, babies receive some of their mothers' flora in utero and through contact with their skin. Plus, breast milk contains oligosaccharides, which act as prebiotics for baby. Prebiotics feed the good gut bacteria and promote bacterial diversity, including IgA antibodies, which help keep bacteria in check, preventing inflammation, infection, and gut problems.

If a mother's microbiome has been disrupted, which affects her immune system, her baby could have an increased lifelong risk of health problems, so if you're thinking of becoming pregnant, take three to six months to focus on restoring your gut health.

What Is Autoimmune Disease?

Autoimmune disease is a state of immune confusion that results in the body destroying its own tissues. Under normal conditions, your immune system is always on the lookout for proteins that are not "you." The immune system gets it wrong sometimes when it also flags proteins that are *definitely* you, marking them for destruction, leading to autoimmune disease. So instead of attacking bacteria, viruses, parasites, and so on, your immune system begins to attack your body. Cue your older sibling: "Quit hitting yourself. Quit hitting yourself."

Your immune system can also target your hormones, creating antibodies— proteins that flag molecules for destruction—against estrogen and progesterone. Normally only foreign molecules should be flagged for destruction, but in autoimmunity, your healthy tissues and hormones are the focus of attack. When this occurs, it can delay or hinder ovulation and prevent the uterine lining from thickening, making it difficult to get pregnant, causing erratic or short periods, or inducing extremely short or long cycles. Insufficient hormones can lead to mood swings, irritability, depression, and anxiety. Your body can also make antibodies to your ovaries, which can lead to early menopause or what is known as primary ovarian insufficiency.

If you suffer from an autoimmune disease, know this: autoimmune diseases are reversible and remission is possible. I've reversed the symptoms of my own autoimmune thyroid and adrenal disease, and I've helped thousands of women with all different autoimmune conditions do the same. The 30-Day Brighten Program can help you counteract the damage and inflammation caused by the pill, heal your gut, balance your hormones, and reduce your risk for autoimmune disease.

What Triggers an Autoimmune Disease?

Autoimmune diseases were barely on the radar a generation ago but are now dramatically on the rise, especially among women, who are ten times more likely than men to have them. Some estimates suggest that approximately one in five people have an autoimmune disease. Gender alone places women at a higher risk (perhaps because so many of us have been on the birth control pill for years).

Let me explain how an autoimmune disease develops so you can understand why the pill can put you at risk. There are essentially three ingredients necessary to produce an autoimmune disease, as defined by Dr. Alessio Fasano, a gastro-enterologist and leading researcher in the field of autoimmunity: leaky gut, genetics, and a stressful event, like environmental toxin exposure or infection.

INTESTINAL HYPERPERMEABILITY (AKA LEAKY GUT)

When your gut barrier breaks down, aka leaky gut, foreign proteins (food and microorganisms) gain access to your bloodstream . . . and your immune system ain't having none of that. Since about 70 to 80 percent of your immune system hangs out in your gut, that's the prime place to set off the immune confusion and turn your white blood cells on yourself.

A GENETIC PREDISPOSITION—OR WHAT YOUR MAMA GAVE YA!

Like a road map, your genes can determine which direction you go, but they don't control how you get there. If, for example, you have a family history of Hashimoto's thyroiditis, you may have the genes for the disease and there-fore be at risk, but without a trigger and a leaky gut, you may never develop Hashimoto's.

A TRIGGERING EVENT

Triggers that challenge your immune system's integrity and increase the like-lihood it will make a mistake include food sensitivities/allergies, bacterial or yeast infections, parasites, viral infections, chronic stress, the pill, antibiotics, and pregnancy.

The Estrogen–Autoimmune Connection

Research has shown that estrogen enhances the inflammatory process of the immune system, meaning it could increase the attack on your tissues if you al-ready have an autoimmune disease, but it can also be a source of the dysfunction.

Your immune system is more robust during reproductive years—when estrogen is at its highest. As you transition into menopause, your estrogen declines and your immune system is more comparable to a man's. For this reason, some women see a decline in their autoimmune symptoms postmeno-pause. Although before the ovaries have called it quits, there are many ups and downs in estrogen production, which can drive autoimmune disease wild. If you have an autoimmune disease, these ups and downs can be experienced as a flare-up in symptoms: sudden joint pain, hair loss, fatigue, changes in skin, and others.

While the decline in estrogen accompanies a decline in immune function, there is also a level of immune dysfunction that can arise. This may be one of the reasons we see more heart disease (which has a strong autoimmune component) in postmenopausal women. It's a bit of a Goldilocks situation—not too much, not too little, but just the right amount of estrogen is necessary for appropriate immune system regulation. For this reason, bioidentical hormone therapy—using hormones that are chemically identical to the natural ones you make, as opposed to the synthetic ones found in birth control—should be considered on an individualized basis. Women should first be encouraged to find and treat the underlying cause of their hormone imbalances, only utilizing medications when absolutely necessary.

Where does the pill fit into all of this? Well, some would say the steady dose of hormone is beneficial because it may prevent the immune flare-ups related to estrogen. However, because the pill is inflammatory, compromises gut lining,

---

Lab Tests for Autoimmune Disease

- antinuclear antibodies (ANA) with reflex—a general screening for autoimmunity
- anti-phospholipid antibodies
- anti–*Saccharomyces cerevisiae* (ASCA) and perinuclear anti-neutrophil cytoplasmic antibodies (p-ANCA)—screens for inflammatory bowel disease
- Cyrex Array 5—predictive autoantibody testing to detect antibodies potentially years before symptoms present
- high-sensitivity C-reactive protein (hs-CRP), a marker of inflammation
- thyroid peroxidase (TPO) and thyroglobulin antibodies—tests for Hashimoto's thyroiditis, the most common autoimmune disease in women
- 21-hydroxylase antibodies—tests for Addison's disease

skews your microbiome, and inhibits healthy adrenal function, it more likely places you at greater risk of a flare-up. Like all things health-related, this is individualized.

Gut Check—Do I Have a Healthy Gut?

Your gut can be vocal when it's unhappy. But even if you're not having symptoms, such as discomfort after meals, indigestion, gassiness, or acid reflux, it's still possible you have gut issues. Use the quiz at the beginning of this chapter to assess your gut health, and consider the list of tests provided on page 114. If you're on the pill, experiencing PBCS, or experiencing any symptoms of a hormone imbalance like headaches, bad moods, acne, and weight gain, then I recommend lab testing to understand the root cause of your symptoms.

Check Your Gut Before You Check the Scale

Leaky gut, gut infections, inflammation, lack of diverse gut flora, and hormonal imbalances lead to weight gain or an inability to lose weight. Not to mention the body will have a harder time absorbing nutrients that are essential for metabolism, and leaky gut is associated with insulin resistance. Add estrogen to the mix and those fat cells get plumper and bigger and start producing more estrogen, as well as leptin. Normally, leptin is a good thing because it signals to your brain that you're full and don't need to keep eating. However, if fat cells keep releasing high levels of leptin, over time your body becomes resistant to these consistently high levels and you experience what is known as leptin resistance, which has the exact opposite effect: instead of feeling full and satiated, you feel hungry, even if you just ate. Unfortunately, leptin is also responsible for perceiving food as a reward, so your increased appetite with leptin resistance is compounded by your feeling of being rewarded when you eat, causing you to eat more. As you consume more and more food, you build more fat cells, which— you guessed it—create more leptin, and the cycle continues.

If you're struggling with weight, it may not be as simple as the "Eat right and exercise" advice we've all heard. In fact, it's generally a lot more complicated when we start talking inflammation, gut health, and hormones. I encourage you to check your gut health before you step on that scale and see if that could be contributing to your inability to lose weight.

Lab Tests for Gut Analysis

Work with a qualified naturopathic or functional medicine doctor to investigate any possible nutrient deficiencies or gut dysfunction. It's important to test for underlying conditions, like bacterial dysbiosis, yeast overgrowth, SIBO, and parasites. Order the following lab tests and keep reading for my tools to restore gut health.

- **Comprehensive Stool Analysis by Doctor's Data or GI Effects by Genova:** Both of these tests will tell you about what's growing—and not growing—in your gut and will also offer insight about your overall digestion.

- **Lactulose or glucose breath test:** If you have a history of food poisoning or regularly experience heartburn, gas, bloating, constipation, or diarrhea, this test can help you get to the bottom of it. This test takes three hours and measures hydrogen and methane gases.

- ***Helicobacter pylori* breath test:** This test will reveal if you have an *H. pylori* infection, a type of bacterial infection that causes stomach ulcers.

- **Celiac testing:** If you are experiencing gut symptoms, you may have an issue with gluten and should consider being tested for celiac disease. Note: you must be currently eating gluten to have accurate results.
  - tissue transglutaminase antibody (tTG-IgA)
  - endomysial antibody (EMA-IgA)

What Is SIBO—and Do I Have It?

Small intestinal bacterial overgrowth, commonly called SIBO, is a condition in which very good bacteria are in the wrong place in your intestines. These bacteria eat the carbohydrates you don't absorb, like fiber. But when they get into your small intestine, they disrupt your bile acid and pancreatic enzymes, which allows them first dibs on your food. SIBO can lead to nutrient depletions, inflammation, leaky gut, brain fog, and a whole host of other symptoms.

- deamidated gliadin peptide antibodies (DGP IgA and IgG)

- total serum IgA

- genetic tests HLA DR3-DQ2 and DR4-DQ8—99 percent of those with celiac disease have either one or both of these genes.

- **IgG and IgA food intolerance:** Food intolerance testing can expose leaky gut and help you understand which foods to avoid in order to heal your gut.

The Brighten Gut Repair Protocol

If you've been on the birth control pill at any point, it has done a number on your gut, as can age, antibiotics, stress, poor dietary choices, alcohol, and a whole slew of other activities you engage in through your lifetime. Remember, your gut health is essential for replenishing the nutrient stores the pill has been depleting, modulating your immune system, keeping inflammation down, helping you have a healthy weight and a healthy mood, and moving out the excess estrogen from your body. For this reason, showing your gut some love can go a long way toward fast-tracking your healing and rebalancing your hormones.

If you're on the pill and need to stick with it, you'll want to do this protocol now and tend to your gut daily. And you'll want to repeat it if and when you decide to discontinue the pill. In my clinic, I have patients repeat a gut-repair protocol at least four times a year while they are on the pill.

If you're coming off the pill, follow this protocol for at least 30 days to support your body's transition and avoid PBCS. Many women coming off the pill benefit from 90 to 120 days of supplements. In chapter 12 I'll show you how to combine this with the Brighten Detox Protocol to accelerate results.

The key steps of this protocol are **NEAT**:

Nourish	Feed your gut the nutrients it needs for optimal health.
Eliminate	Remove aggravating foods from your diet.
Absorb	Support digestive function to absorb nutrients.
Terrain	Optimize the environment of your gut and all the critters it houses.

Nourish

When you nourish your gut, you nourish your body. Remember, the pill depletes important nutrients that your body needs, such as antioxidants, B vitamins, selenium, and zinc. The 30-Day Brighten Program dietary approach has two components: introduction of foods that build healthy hormones and removal of foods that may be aggravating your symptoms. Leveraging food as medicine restores hormone balance and undoes the negative effects of the pill. In addition to the foods recommended in chapter 5 on page 97, begin consuming these:

- **Bone broth:** Bone broth is one of the most healing foods you can ingest when you're trying to heal a leaky gut. It consists of vegetables, herbs, and the bones of animals, which contain vitamins, minerals, and amino acids that are therapeutic for gut healing. Bone broth also has large amounts of the amino acids glycine and proline, key components of connective tissue and critical in healing the microscopic damage caused by the pill. With its high levels of glycine, bone broth can help regulate the immune system and reduce inflammation too. The amino acids in bone broth are easily assimilated by the body, making it a perfect choice when you're working to heal your gut. Make your own bone broth or purchase a high-quality version from a local store that does not contain any harmful additives and is made with bones from healthy, grass-fed animals.

- **Upgraded Golden Milk:** Drink this nightly for healthy motility and lower inflammation. (See page 304 for the recipe.)

- **Gut-promoting foods:** Sauerkraut, dill pickles, water kefir, coconut yogurt, and other fermented foods naturally contain probiotics, as well as the fibers your gut bugs need to thrive. These foods nourish the health and function of the microbiome, which in turn supports the health of your gut.

L-GLUTAMINE

Supplementing with this amino acid has been shown to improve intestinal cell health and reduce leaky gut associated with chronic medication use. The cells of your intestines turn over rapidly, about every three to six days, and L-glutamine, along with other essential nutrients, can aid in gut healing. Typical adult dosing is about 1.5 to 5 grams two to three times daily for thirty to ninety days. If you stick with the pill, consider taking 1 to 2 grams daily ongoing or 5 to 15 grams daily for thirty days every three to four months.

OMEGA-3 FATTY ACIDS

Omega-3s have been shown to be beneficial in reducing inflammation, supporting a healthy mood and brain function, and improving gut health. Omega-3s include alpha-linolenic acid (ALA), eicosapentaenoic acid (EPA), and docosahexaenoic acid (DHA). Mackerel, salmon, sardines, and other cold-water fish are excellent sources of EPA and DHA. ALA is found in whole flaxseeds, walnuts, and chia seeds. ALA can be used to synthesize EPA and DHA, although this is highly inefficient in humans and appears to be dependent on your natural estrogens, which is why I recommend eating animal sources of omega-3s or supplementing with cod liver oil. Algae is a source of DHA, which can be converted to EPA, making it an option for vegans. But for women with PBCS or on the pill, I recommend food and supplement sources that are higher in EPA and DHA when possible. Most people's diets are high in pro-inflammatory forms of omega-6, like canola, corn, cottonseed, grapeseed, peanut, safflower, and soybean oil, along with other processed foods. Omega-6s, like omega-3s, are essential to health, but the trouble is they are often out of balance when our diet favors omega-6s. Instead, aim to get your omega-6 from sources like pine nuts, sunflower seeds, Brazil nuts, and pecans. And start including more omega-3-rich foods in your diet.

Most women experiencing gut and hormone symptoms do best supplementing with 1,000 to 4,000 milligrams of combined EPA and DHA daily. If you're on the pill, I recommend 1,000 to 2,000 milligrams of EPA and 500 to 1,000 milligrams of DHA ongoing.

N-ACETYLCYSTEINE (NAC)

NAC is the amino acid your body uses to make glutathione, a potent antioxidant that protects your cells. NAC supports liver detoxification, aids in gut healing, breaks down biofilms (think force fields that bad gut bugs make), and prevents free-radical damage. NAC is typically dosed at 600 to 1,800 milligrams daily for three to six months. When we're talking fertility (and we will—see chapter 10), then women typically stay on this dose through the first trimester, but of course always check with your midwife or doctor first.

ZINC

For such a little mineral, zinc plays a big role in your body. From immune function to wound healing to thyroid function and a little bow-chicka-wow-wow, this mineral is necessary for it all. It's also an essential nutrient in healing your intestines, and remember, zinc is one of those nutrients the pill depletes.

Deficiencies in zinc are associated with an increase in gut permeability. Zinc has been shown to be beneficial in healing intestinal permeability in patients with Crohn's, an autoimmune condition that pill users are at increased risk for. It also has been shown to be helpful in increasing intestinal integrity, aiding in gut health, and reducing the duration of bacterial gut illnesses.

Signs of a zinc deficiency include thin, brittle nails that may have white spots, a tendency to get sick easily, and difficulty smelling or tasting. Supplementing with zinc as part of a multivitamin or prenatal can help. Typical doses are about 15 to 30 milligrams daily, although doses as high as 75 milligrams have been used in studies to repair the intestines. Doses of zinc above 40 milligrams per day can increase proteins that prevent copper absorption and create a deficiency, causing anemia that is not responsive to iron and low neutrophils, a type of white blood cell.

Reasons you may not be getting enough zinc include the pill, celiac disease, iron supplements, poor intestinal health, low stomach acid, and a diet high in phytates, like legumes and grains. There also have been studies showing that alcohol inhibits the absorption of several nutrients, including zinc. That's another reason you gotta ditch the alcohol to heal your gut and your hormones.

VITAMIN D

Vitamin D is necessary for healthy immune function and hormone health. Keep in mind that we primarily synthesize vitamin D in our skin in response to sun exposure. If you're deficient (less than 30 ng/mL), that's a definite sign you're in need of a little more time outside in nature and in the sun! But it also may be a sign of gallbladder dysfunction, pancreatic issues, SIBO, or other causes of fat malabsorption. A blood test can tell you if you're deficient. Optimal levels are between 60 and 80 ng/mL. Some studies suggest that oral contraceptives may increase levels of vitamin D, which can lead to cardiovascular complications. When women stop oral contraceptives, their vitamin D levels may drop. This is why it's important when discontinuing the pill to monitor with lab work while supplementing.

DEGLYCYRRHIZINATED LICORICE (DGL)

DGL is a form of licorice that helps support the mucosal lining of the stomach and the intestines without affecting hormones. Some people find it beneficial for heartburn relief. I recommend 500 milligrams one to two times daily.

SLIPPERY ELM, CHAMOMILE, AND MARSHMALLOW ROOT

Teas, tinctures, and supplements of slippery elm, chamomile, and marshmallow root can be very healing and soothing to the intestines plus beneficial when you're recovering from leaky gut. Slippery elm can be made into a gruel.

Eliminate

Food sensitivities and leaky gut go hand in hand. Because food proteins can make their way through the space between your cells, aggravate your immune system, and create inflammation, you'll need to eliminate some key foods while you heal your gut. Migraines, difficulty losing weight, acne, hormonal imbalances, fatigue, and many other symptoms can be signs of food sensitivities. This is much different from a food allergy, which is mediated by the immunoglobulin E (IgE) aspect of the immune system and results in hives, swelling, and difficulty breathing. Because food sensitivities can be difficult to identify (a reaction may not be evident for up to three days after consuming the food), many people have no idea they are reacting to a food.

Why is it important to determine if you have food sensitivities? Consistently eating foods that you're sensitive to constantly triggers your immune system to cause damage within your body and increase inflammation. You may find that despite your best efforts to feel better, you just can't achieve the state of health you're after. As I have seen, by uncovering food sensitivities and implementing dietary changes, patients improve dramatically in their symptoms and even reverse their chronic conditions.

If you'll recall, the top inflammatory foods you'll need to eliminate are gluten and grains, dairy, soy, sugar, caffeine, alcohol, and inflammatory fats. Avoiding these foods for 30 days will give your gut time to heal from their effects. After 30 days, you'll gradually reintroduce them one at a time and a few days apart to determine if they give you a reaction. For example, if you reintroduce dairy and find that you develop acne, then you know that you and dairy will have to part ways. (For specific reintroduction information, see chapter 12.) What about eggs and shellfish? While a traditional elimination diet often suggests you remove these two foods, clinically I've observed that this creates a great deal more stress for women who are also working to undo the damage caused by the pill. For the purpose of this program, you can continue these foods, unless you know you have an allergy or a sensitivity to them.

GLUTEN AND GRAINS

Gluten is a protein found in many grains—wheat, barley, rye, spelt, and others. It's a common food sensitivity and can contribute to leaky gut and increased inflammation as well as aggravate certain autoimmune conditions. Because gluten is found in a lot of popular foods, you'll need to read labels and double-check that you are not accidently exposed.

DAIRY

Dairy, also a common food sensitivity, causes inflammation, which can sometimes be reversed once the gut is healed. If you're experiencing acne and skin symptoms, dairy may be aggravating them further. You'll need to avoid all forms of dairy, including butter, cheese, whey protein, and ice cream. If you're wondering where you'll get your calcium, then rest assured. You'll have plenty of rich and bioavailable sources of calcium throughout the meal plans in chapter 12.

Dairy Alternatives

Dairy Item	Replacements
Butter	Coconut oil, coconut butter, olive oil, camel hump fat, lard
Cheese	Nut cheese, nutritional yeast
Ice cream	Almond ice cream, cashew ice cream, coconut ice cream
Milk	Coconut milk, cashew milk, almond milk, camel's milk
Whey protein	Hydrolyzed beef protein, pea protein, cricket meal, bone broth protein
Yogurt	Almond yogurt, cashew yogurt, coconut yogurt

SOY

Soy disrupts hormones . . . and it's everywhere! Besides the obvious soy sauce, miso, and edamame, soy can be in everything from bread, cookies, and other baked goods to soups and sauces. (Again, read your labels!) Soy can minimize the effects of your natural estrogen, lower your fertility, increase inflammation, and lead to issues with adrenal and thyroid function. How do you know if this is true for you? Eliminate and reintroduce.

Five Tools to Help You Quit Gluten

1. **Know where gluten hides.** You may know that gluten is in your bread, pasta, cake, and cookies, but did you know that it's also potentially in your vitamins, herbal teas, makeup, shampoo, and over-the-counter drugs? On your supplement and medication labels, look for the ingredient "dextrin," a wheat-derived filler.

2. **Read labels.** Be sure to read all the labels on your food. Gluten is commonly found in vegetarian products, soy sauce, meatballs, condiments, and even soups. Better yet, try to consume only whole foods, like meat, fish, and vegetables, which do not come packaged in a box with twenty other ingredients.

3. **Don't be afraid to ask when you're dining out.** Be sure to ask your server if whatever you plan to order contains gluten. You never know when flour or breadcrumbs may be added to a dish as a thickening agent. Many restaurants are beginning to cater to their gluten-free clients, and they understand the severity of food allergies.

4. **When in doubt, skip it.** If you really aren't sure, avoid a food that could be contaminated with gluten. It's not worth the risk.

5. **Remember BROWS.** Many different grains contain gluten, including barley, rye, oats, wheat, and spelt (BROWS). While oats do not naturally contain gluten, they are often manufactured on equipment that also produces gluten-filled grains, so there is a risk of cross-contamination.

SUGAR

Sugar causes inflammation in your body, which of course leads to leaky gut and insulin resistance. Because of the way the birth control pill affects insulin and adrenal function, you may actually feel increased sugar cravings while on it. I won't lie: giving up sugar is probably going to be tougher than eliminating the other foods in this section. First of all, sugar is in *everything*. It's even in chicken broth, ketchup, and salad dressing. So, if you don't eat a lot of sweets, you may still be consuming far more sugar than you realize. Eliminating sugar includes artificial sweeteners, like the kind found in sugar-free and diet foods. Also, watch out for fruit juice and other sugary drinks.

During the first few days of no sugar, you may experience a bit of withdrawal, with symptoms like headaches, fatigue, and irritability. But if you can get past

---

Caffeine-Free Coffee Replacements

The following beverages are great substitutes for that morning—or afternoon!—cup of coffee, and they're also healthy for you.

- **Maca Latte** (see recipe page 305)

- **Upgraded Golden Milk** (see recipe page 304)

- **Mushroom elixirs** have a variety of health-boosting effects, and many companies, such as Four Sigmatic and SuperFeast, make instant elixirs. (Be aware that some "mushroom coffee" options are still mixed with coffee; get one that is exclusively mushroom based.)

- **Decaf matcha**—green tea powder—is typically whisked into hot water or another nondairy liquid: ½ to 1 teaspoon per 8 ounces of liquid.

- **Chicory root and dandelion root,** when roasted, take on a coffee-like flavor. Use ½ teaspoon each of chicory root and dandelion root per 8 ounces of water. Steep for 3 to 5 minutes, then strain.

- **Herbal teas**

- **Spa water**—add cucumber, berries, lemon, or lime to water for added flavor.

---

those first few days, you will be over the hump, and it will get much, much easier.

CAFFEINE

This is temporary. Whew! Okay, now that we've got that out of the way, let's dive into why you're going to cut it out. While it's true that coffee has many health benefits, caffeine can be too stimulating for the nervous system and can interrupt sleep or trigger hot flashes. If you're worried about going cold turkey, then I recommend weaning yourself off gradually. Try reducing your coffee intake by 50 percent every three days until you're no longer drinking coffee. For example, if you currently drink a cup of coffee every day, try a half-caf cup for three days, then drink only a quarter cup of caffeinated coffee for three days, then have none for the remaining duration of the program. It's best not to simply continue drinking decaf coffee, because it can be irritating to your gut.

ALCOHOL

Alcohol can cause leaky gut, blood sugar imbalances, and bacterial overgrowth in your small intestine. If you find yourself missing a fun beverage during this protocol, then try the Anti-inflammatory Turmeric Spritzer (see page 304).

What Foods Are In or Out?

In	Out
Healthy fats (avocado, coconut oil, cold-water fish, macadamia nut oil, olive oil)	Inflammatory fats (trans fats, processed foods, fast foods, canola oil, corn oil, cottonseed oil, peanut oil)
Organic fruits and vegetables	Gluten and grains
High-quality protein	Soy
Water, Upgraded Golden Milk (see page 304), other nourishing beverages	Alcohol and caffeine
Bone broth	Dairy
Fermented foods	Sugar and artificial sweeteners

INFLAMMATORY FATS

Fats help regulate your hormones, but as with many other things in life, quality matters. There are good fats and not-so-good fats. Follow the 30-Day Brighten Program meal plans to get plenty of the healthy fats you need, like essential fatty acids. In order to heal the gut, however, eliminate inflammatory fats, like the trans fats in processed foods, margarine, and fast foods.

Absorb

You aren't what you eat—you are what you eat and absorb. You must support the digestive process of breaking down nutrients in order to absorb them. You're already beginning to do this by nourishing your gut, eliminating aggravating foods, and resetting your terrain. The following steps will also improve your digestion:

- **Take HCl.** Your stomach produces hydrochloric acid (HCl) to aid with digestion, but HCl is also a powerful antimicrobial. It helps to free the nutrients from your food, like iron, vitamin B12, calcium, and protein, to allow your small intestine to absorb them. When you have an optimal hydrochloric acid level, other digestive processes operate more effectively and efficiently. Taking betaine HCl with pepsin can support the digestive track overall.

- **Add digestive enzymes.** A combined supplement of amylase, pepsin, lipase, and protease can help you make the most of the food you're eating. If you're on the pill and want your body to absorb food more effectively, leverage digestive enzymes too.

- **Try ox bile.** Support your body's absorption of fat and fat-soluble vitamins, and give your gallbladder some much-needed assistance, with 75 to 500 milligrams daily of ox bile.

- **Include natural prokinetics.** If you're experiencing any constipation, you may want to add ginger, magnesium, or vitamin C supplements to help with intestinal motility and moving out waste.

- **Reduce stress.** Stress or sympathetic dominance can alter your microbiome and inhibit blood flow to your gut. When you're constantly stressed and in a sympathetic ("fight or flight") state, it dramatically affects your digestion. If this type of stress is an issue, try eating mindfully, enjoying all the sensations of your meal, chewing thoroughly,

Lifestyle Strategies to Support Your Gut

- Eat in a relaxed environment.

- Practice mindfulness eating. Be present with your food.

- Avoid drinking lots of fluids with your meals.

- Reduce stress by singing out loud (it's good for your vagus nerve too, which supports healthy gut motility)

- Avoid moderate to intense activity an hour after eating. Let your body *rest and digest*.

- Practice deep breathing as often as possible.

- Indulge in castor oil belly massages with peppermint essential oil.

and not working while you eat. Practicing relaxation techniques, such as meditation or yoga (see "Stress-Reducing Practices," page 260), and knowing and setting your limits are as important as eating veggies.

- **Remove unnecessary medications.** Stop taking NSAIDs, proton pump inhibitors (PPIs), unnecessary antibiotics, and birth control pills (if you're ready). And of course, if you doctor prescribed it, have a chat with them first.

Terrain

Imbalances in the microbiome, like those created by the pill, leave the gut susceptible to infection. The pill also induces inflammation, which makes for a seriously hostile environment for those beneficial bugs and hinders your ability to break down and absorb nutrients. This is especially problematic if you've ever been on the pill, which is known to deplete the body's nutrients. Take the following steps to rebalance your microbiome and make a happy home for the critters so you can begin to support your gut and hormonal health, and heal from the effects of the pill.

ADD FERMENTED FOODS TO YOUR DIET

Fermented foods like beet kvass, kombucha, sauerkraut, and kimchi are filled with beneficial microbes and prebiotic fibers that can help restore gut health. While eliminating dairy, include nondairy fermented alternatives in your diet, like water kefir and coconut yogurt. Many of my patients benefit from fermented foods, although, if you have SIBO, fermented foods and certain probiotics can aggravate symptoms and will need to be avoided until the infection is cleared.

AVOID PESTICIDES

Foods laden with pesticides and antibiotics cause disruption to the microbiota or simply kill off the good guys. As often as possible, choose organic, locally grown food.

BALANCE GUT FLORA WITH PROBIOTICS

What I've seen clinically is that a high-quality probiotic is absolutely essential for healing the gut as women come off the pill. If you choose to stay on the pill, then you should take a probiotic daily and rotate types every three to four months to introduce a variety of strains. The organisms I have seen be most helpful for women who have been on the pill are high-dose, high-strain *Lactobacillus* and *Bifidobacterium* species, *Saccharomyces boulardii,* and spore-forming probiotics. When choosing a probiotic, quality matters. I recommend being very picky when it comes to all your supplements, and probiotics are no exception.

When starting probiotics, go slow to avoid symptoms. See the Brighten Supplement Protocol (page 246) for my recommendations. When selecting a high-dose, high-strain *Lactobacillus*- and *Bifido*-based probiotic, you want to look for at least 10 strains. These bacteria can help with immune support and provide you with beneficial flora that help synthesize vitamin K and B vitamins that are depleted by the pill.

I also recommend using spore-forming probiotics, which can help boost the *Lactobacillus* species in the gut when taken with a *Lactobacillus* probiotic. If you're concerned you have SIBO, spore-based probiotics can actually help eliminate SIBO, and unlike the *Lactobacillus* species, they're less likely to cause any kind of discomfort. Research has also shown that spore-based probiotics, unlike other probiotics, survive in stomach acid and may help you repopulate your gut.

When you start spore-based probiotics, I recommend taking a half cap each day for seven days. You can open up a capsule and empty it into your mouth or onto food. After seven days, increase your dose to one cap daily for fourteen days. After fourteen days, increase to one cap twice daily for fourteen days, after which you can increase to two caps twice daily for at least sixty if not ninety days.

Saccharomyces boulardii is a beneficial yeast that can increase your secretory IgA to a healthy level. Secretory IgA is an antibody that lines the intestines and protects the body from foreign invaders, like parasites and opportunistic pathogenic infections. Chronic stress and inflammation, like what can be experienced while taking the birth control pill, can deplete this antibody, making you vulnerable to infections.

Note that when taking probiotics, "die-off" can occur in your body, commonly known as the Herxheimer, or Herx, reaction. When this occurs, you may experience flu-like symptoms, such as headaches, body aches, an upset stomach, or rashes. While you may feel lousy, it's often a sign that healing is taking place as you eliminate bad bacteria in favor of good bacteria. But when in doubt, see your doctor.

See the Brighten Supplement Protocol, page 246, for recommended probiotic strains to incorporate into your regimen and the recommended doses during this program. And for the most up-to-date probiotics choices, please visit DrBrighten.com/Resources.

Key Takeaways: Gut Check

- Your female hormones, your gut, and your microbiome are intimately connected.

- A healthy gut is essential for removing excess estrogen and avoiding a hormonal imbalance.

- The pill alters your microbiome and destroys your gut integrity; it also increases your risk of autoimmune disease.

- Leaky gut occurs when the tight junctions between the cells of your intestinal lining are compromised and can no longer protect you from foreign invaders, leading to inflammation and hormonal disruption.

- The top causes of leaky gut include NSAIDs, antibiotics, the birth control pill, chronic stress, high-sugar diets, and low-fiber diets.

- Gut imbalances may be the cause of your weight gain.

- Autoimmune disease develops when you have leaky gut, a stressful event, and a genetic predisposition.

- The Brighten Gut Repair Protocol includes the following four key steps: nourish, eliminate, absorb, terrain (NEAT).

- L-glutamine, omega-3 fatty acids, NAC, zinc, vitamin D, DGL, slippery elm, chamomile, marshmallow root, probiotics, and digestive enzymes can help heal the gut.

CHAPTER 7

ENERGIZE YOUR THYROID AND ADRENALS

Do you have a thyroid or adrenal condition? Then I've gotta tell you that the pill is doing you no favors. Between inflammation, nutrient depletions, and increasing cortisol and thyroid-binding proteins, the pill disrupts thyroid and adrenal function in a major way. You see, the thyroid contributes to our mood, our metabolism, our menses, and our energy, and it affects every single cell in the body. Without adequate thyroid hormone, we feel depressed and have brain fog, plus our skin becomes super dry, our hair starts to fall out, and we feel achy and older than we should. And we often gain weight and are unable to tolerate exercise.

I need to share some facts with you and give you a little reality check on how important your thyroid function is. The American Thyroid Association estimates that more than 27 million Americans have thyroid disease. But unfortunately, more than half of them are walking around with no idea they have a thyroid condition. Women are five to eight times more likely than men to develop a thyroid disease. In the United States, the autoimmune disease Hashimoto's thyroiditis, or autoimmune thyroiditis, is the number one cause of thyroid disease. That means the body's immune system attacks its own thyroid gland. While research is working hard to understand just why we ladies are so susceptible to this autoimmune disease, it's pretty clear our hormones are in play. But let's not forget that about 80 percent of women have taken the pill, which is inflammatory, contributes to immune dysfunction, and causes leaky gut.

Your thyroid produces thyroxine hormone, or T4, which must be converted into triiodothyronine hormone, or T3, for your cells to be able to use it. Your gut is one of the major thyroid hormone conversion sites, which means it takes the inactive hormone and activates it so you can feel full of energy and happy throughout your day. But if your gut is inflamed, your microbiome is skewed, you're being depleted of nutrients, and/or you have leaky gut, that's going to impact the activation process. And without active thyroid hormone, your gut can't function optimally. This is where women get stuck in a vicious cycle, because if thyroid hormone levels fall too low, the body doesn't produce enough hydrochloric acid, the gallbladder doesn't work properly, and the gut doesn't move (read: constipation). Now, if you're starting to freak out a little bit and thinking that you're a wreck, no worries—I've got you in this chapter. I'm going to break down how the pill affects your thyroid and how to restore your thyroid health.

The 411 on Your Thyroid Hormones

Thyroid stimulating hormone (TSH) is how your brain speaks to your thyroid. Your thyroid then responds by secreting T4 and a tiny bit of T3. T4 is the inactive hormone that travels to other tissues in your body, namely the gut, liver, and kidneys, where it gets converted into T3, your active thyroid hormone. As I already mentioned, T3 affects your mood, energy, menses, and metabolism, plus it keeps you warm. (If you're always cold, you may want to have your thyroid checked.) Anything that hinders the production of TSH or T4, or the conversion of T4 to T3, or blocks your cells from using T3 can cause hypothyroid symptoms. A number of things can disrupt your thyroid: heavy metals, fluoride, infections, inflammation, autoimmune disease, stress, nutrient deficiencies, and medications, including the birth control pill. While many thyroid conditions are autoimmune, we still have to determine what the root cause is, because someone with a thyroid autoimmune disease may also have nutrient depletions or any of the other issues just listed.

There are two primary ways thyroid disease manifests, as either hyperthyroidism or hypothyroidism. Hyperthyroidism is when you have an overactive thyroid making too much thyroid hormone. Conversely, when you have an

In This Chapter

- How the pill sabotages your thyroid and adrenals

- The many ways thyroid disease can present

- The thyroid–gut connection (Hint: leaky gut can put you at risk for hypothyroidism.)

- How to support your adrenals right now

- How to improve your thyroid health on or off the pill

- How your sex hormones depend on adrenal and thyroid function

underactive thyroid not producing enough thyroid hormone, it's hypothyroidism, which is much more common than hyperthyroidism.

Ignored Thyroid Symptoms

While you may suspect you have a thyroid disease based on the symptoms listed in the table on page 132, there are some more subtle ways thyroid disease presents that you don't want to miss:

- **Do you suffer from period problems and PMS?** Irregular cycles, spotting between periods, heavy bleeding, cramps, and/or mood swings may be signs of thyroid disease. If you have a hormonal imbalance, you may have too much estrogen, which can enhance the inflammatory process of your immune system and contribute to the attack on your thyroid.

- **Did you have major fatigue after starting the birth control pill?** The pill causes an increase in thyroid-binding globulin, which binds your free thyroid hormone so that you can't use it. We'll explore how the pill disrupts your thyroid a bit later in the chapter.

- **Has your sense of smell or taste changed?** If you aren't producing enough thyroid hormone, it can alter the way you smell, taste, and therefore enjoy your food. If you find yourself becoming less interested in food and losing your appetite, you may have hypothyroidism. The good news is that once you treat the condition, these senses return.

- **Are you experiencing vision changes and eye disorders?** Hashimoto's thyroiditis and Graves' disease can cause eye disorders. When you have episodes of elevated thyroid hormone, which can happen with these diseases, it can stimulate eye growth, causing your eyes to protrude or stick out, which can damage your cornea. If your eyes are often itchy, you experience a delay when you try to close your eyelids, or you notice changes in your vision, schedule a visit with your eye doctor to be evaluated properly.

- **Are you grappling with major exhaustion during the first trimester of pregnancy?** The pregnancy hormone hCG (human chorionic gonadotropin) stimulates thyroid production during the first half of pregnancy. If you're pregnant and struggling with

Symptoms of Thyroid Disease

Hyperthyroidism	Hypothyroidism
Weight loss	Weight gain
Heat intolerance or sweating excessively	Cold intolerance
Anxiety	Depression and/or anxiety
Racing heart	Slow heart rate
Shaky hands	Delayed reflexes
Insomnia	Fatigue and memory loss
Hair loss	Thinning hair, dry skin, brittle nails
Loose stools	Constipation
Menstrual irregularities	Menstrual irregularities

fatigue, constipation, dry skin, hair loss, or depression, it may be hypothyroidism. I highly recommend having your thyroid checked before you become pregnant, early in your first trimester, and postpartum. If your TSH is 2.5 µIU/mL or higher, then you need to consider taking a thyroid medication for your health and your baby's. Postpartum thyroid disease affects one in twelve women worldwide.

- **Do you have SIBO?** SIBO is actually really common in thyroid patients. If you have gas, bloating, constipation, diarrhea, or other gut symptoms, it may be due to hypothyroidism.

- **Do you have serious heartburn?** Heartburn is that burning sensation in your chest that you may experience after meals or even throughout the day. You need thyroid hormone to make stomach acid, which means that without it you can't properly break down food to get all the nutrients. In fact, your digestion and thyroid are intimately connected. When you have adequate stomach acid, your esophageal sphincter closes. When you have low stomach acid, it may not receive the signal to close and then acid can make its way into your throat, causing heartburn.

- **Is your voice changing?** Does it sound like you smoke a pack of cigarettes when you get up in the morning? That can be a sign of hypothyroidism. Thyroid hormone prevents a buildup on the vocal cords of sugars called mucopolysaccharides, so when thyroid hormone levels drop, these sugars accumulate on the vocal cords, causing a deeper voice. The vocal cords can also become swollen, which is another cause of voice changes and a common symptom of hypothyroidism. Is your voice feeling shaky and unsteady? That too could be a sign of hyperthyroidism. This is due to the hypermetabolic state—everything is in supercharged mode.

- **Do you have hand or foot pain?** Carpal tunnel can be a sign of thyroid disease. Low thyroid hormone impairs nerve function and blood flow, resulting in cold hands and feet. You need thyroid hormone to repair tissues, which means the muscles that get used the most (the hands and feet) can be the first to show signs of low thyroid hormone, generally manifesting as aching or burning pain. If you're hyperthyroid, then your hands or other muscles may often feel weak. Muscle cramps can occur with both hyper- and hypothyroidism, which is why lab testing is important.

If you answered yes to any of these symptoms, take the following quiz and ask your doctor for a complete thyroid panel, along with antibodies. Correct testing is crucial to diagnose and ensure proper treatment of thyroid disease. Test results should also be interpreted in conjunction with your symptoms. Thyroid labs are one snapshot in time, which is why tracking your symptoms is an important part of the data.

How the Pill Sabotages Your Thyroid

There is a very real connection between the pill and your thyroid. In fact, many of my patients notice that their thyroid problems began after they started taking birth control.

First, the birth control pill depletes vital nutrients that your thyroid requires to synthesize thyroid hormone, as well as the nutrients your cells need to utilize your hormones. Among the nutrients necessary for making and modulating thyroid hormones are the following:

- **Selenium and zinc** are needed to produce thyroid hormone and to convert it to its active form, T3, from the inactive form T4.

- **Zinc** is also required to get thyroid hormone and cell receptors "talking." Just by depleting zinc alone, the pill can prevent you from making, activating, and using thyroid hormones.

- **B vitamins** are key to the synthesizing of thyroid hormone, among hundreds of other bodily functions.

All of these depletions also interfere with your body's ability to use thyroid hormone at the cellular level. If you have a thyroid condition and take the pill, you will make less thyroid hormone, convert less to its active form, and use less of it, which is enough of a reason not to take the pill. If you've been on the pill or are on the pill right now, then your first step to protect thyroid function is to take a quality prenatal or multivitamin to replenish these crucial thyroid nutrients.

Second, the pill increases thyroid-binding globulin (TBG). Elevated TBG causes free thyroid hormone to become bound—so now any thyroid hormone your body can manage to make is unavailable to your cells. Some studies have shown the pill actually increases thyroid hormone, but it's important to note that they were reporting on total (bound) thyroid hormone, which you can't use.

Third, the pill is inflammatory, which is important because inflammation is at the root of all chronic disease. If you already have an autoimmune condition, such as Hashimoto's thyroiditis, more inflammation will only exacerbate your symptoms, making it harder to move, to concentrate, to have energy—to live your best life. When inflammation is high, your body converts your T4 to reverse T3 (RT3), which I call the hibernation hormone because elevated RT3 is designed to make you store calories (aka fat) and want to go to sleep. Inflammation also makes cell walls less responsive to *all* hormones, including thyroid, insulin, and progesterone—to name a few. Insulin resistance leads not only to diabetes but also to neurological issues and heart disease (more on this in the next chapter).

Think of inflammation as fire and the pill as gasoline. Women taking the pill experience an elevation in high-sensitivity C-reactive protein (hs-CRP), which indicates increased inflammation in the body and is associated with a higher risk of chronic disease and mortality. Other markers for acute inflammation, including the proteins fibrinogen and ceruloplasmin, also tend to be elevated with the pill. And as we'll explore in chapter 9, all that inflammation is really bad for the brain.

TSH Is *Not* Enough!

In addition to measuring thyroid-stimulating hormone (TSH), your doctor should test the following:*

- total T4
- total T3
- free T4
- free T3
- reverse T3
- anti-thyroperoxidase (anti-TPO)
- anti-thyroglobulin antibodies
- thyroid receptor and thyroid stimulating immunoglobulins (if there are signs of hyperthyroidism)

*Biotin in doses of 5 to 10 milligrams (which is common in hair-loss formulas) can interfere with test results and should be stopped two days prior to testing.

Quiz: How Is My Thyroid Health?

Check any boxes that apply to you in the following lists:

Part I

☐ I have unexplained weight gain.

☐ I always feel cold.

☐ I have constant fatigue.

☐ I'm depressed or anxious.

☐ I have a slow heart rate.

☐ I have delayed reflexes.

☐ I struggle with brain fog and memory issues.

☐ I have thinning hair, dry skin, and brittle nails.

☐ I am often constipated.

☐ I've lost my sense of smell or taste.

☐ I've noticed changes in my vision.

☐ I'm extremely tired in the first trimester of pregnancy.

☐ I have small intestinal bacterial overgrowth (SIBO).

☐ I get heartburn often.

☐ I have long, painful, or irregular periods—or PMS.

☐ I became fatigued after starting the birth control pill.

☐ I have joint pain or muscle aches.

☐ My face is puffy.

☐ My voice is deep or hoarse.

☐ I feel sore for many days after exercise or physical activity.

☐ My wounds heal slowly.

☐ I have high cholesterol.

TOTAL _____

Quiz: How Is My Thyroid Health?

Part II

☐ I have had unexplained weight loss.

☐ I'm always hot and sweaty.

☐ I have anxiety.

☐ My heart often races.

☐ My hands shake.

☐ I struggle with insomnia.

☐ I'm losing my hair.

☐ I have loose stools or diarrhea.

☐ I'm noticing changes in my vision, my eyes protrude, or my eyes are itchy.

☐ I have menstrual irregularities.

TOTAL _____

Part I:

1 box checked = low risk of hypothyroidism

2 boxes checked = intermediate risk of hypothyroidism

3 or more boxes checked = high risk of hypothyroidism

Part II:

1 box checked = low risk of hyperthyroidism

2 boxes checked = intermediate risk of hyperthyroidism

3 or more boxes checked = high risk of hyperthyroidism

Follow the Brighten Thyroid and Adrenal Health Protocol on page 144 if you have symptoms of a thyroid disorder. And meet with your doctor to have testing.

Dr. Izabella Wentz, a medication safety pharmacist and author of *Hashimoto's Protocol,* notes, "Another trigger relating to some oral contraceptives is the fillers that can be found in the pills. For example, many contain lactose. This may be an additional issue for women with Hashimoto's (who often present with dairy and gluten intolerance issues)."

The good news? Hypothyroidism can be reversed by supporting your thyroid gland and the body's ability to use the thyroid hormone.

The Thyroid–Gut Connection

If you've seen your endocrinologist for hypothyroidism, you were probably given a prescription and sent on your merry way. Did you have a conversation about your gut health? Probably not. But thyroid health and gut health are intimately connected—and you can't heal your thyroid without also healing your gut.

As we discussed in chapter 6, long-term use of the pill can damage the intestinal lining and cause leaky gut. When the gut is compromised, which includes potential infections, the risk of autoimmune disease rises and nutrient malabsorption can occur (remember, the pill is also depleting nutrients). Your body must be nourished to maintain thyroid health, and when it isn't, problems arise. If your thyroid is healthy but your gut is not functioning optimally, you can experience symptoms of hypothyroidism because inflammation inhibits active thyroid hormone (T3). *About 20 percent of thyroid hormone conversion takes place in the gut.* In addition, if you're inflamed, you can't use thyroid hormones at the cellular level. Clinically, I have found that women on the 30-Day Brighten Program who remain on the pill generally experience about an 80 percent improvement in symptoms. However, women who discontinue the pill altogether often report a complete reversal of symptoms with the protocol.

The OAT Axis

Besides your gut, your thyroid is also connected to your adrenals and ovaries. The ovarian-adrenal-thyroid (OAT) axis is the way in which these three hormone-producing glands communicate and truly rule your world. Now, keep in mind, the pill diminishes the communication to and from your ovaries and instead replaces the role of the ovaries with high-dose synthetic hormones. I like to think of the OAT axis as a three-legged stool, with each gland representing a leg. When any one leg is imbalanced, the entire system is affected and begins to falter.

All women depend on proper OAT function to feel their best, and I'm here to tell you, *it's never just one gland that is imbalanced.* When the thyroid, adrenals, and ovaries stop communicating properly or start misfiring, a whole array of symptoms follows.

The Adrenal–Thyroid–Autoimmune Connection

Your adrenal glands play a major role in immune health by responding to inflammation and regulating immune cells through the hormone cortisol. The hypothalamic-pituitary-adrenal (HPA) axis is the mechanism by which your body regulates your cortisol output (that is, how your adrenals and brain are talking) and is affected by inflammation. Your body tries to keep up with the inflammation caused by the pill, but over time there can be a disruption in your HPA axis. This type of disruption, commonly known as "adrenal fatigue," has been shown to accompany a rise in inflammation and autoimmune disease, including autoimmune thyroid disease (Hashimoto's thyroiditis and Graves' disease), multiple sclerosis, Sjögren's syndrome, alopecia areata (hair loss), inflammatory bowel disease (Crohn's and ulcerative colitis), and many more.

As you can imagine, your adrenal glands don't love the inflammatory state caused by the synthetic hormones in the pill. Chronic stress, whether it's mental, physical, emotional, or inflammatory, can disrupt the way your brain and adrenal glands communicate. The problem is that the adrenal glands continue to secrete cortisol whenever they perceive stress, but in today's world, stress can be relentless. City lights at night, driving in traffic, eating on the run, and keeping up with innovations and technology are all subtle sources of constant stress in your life. It's what we call an evolutionary mismatch—your body hasn't had the chance on the genetic level to change its physiology to match our modern environment. It was designed to respond to the occasional stress encountered during our hunting-and-gathering days, when we had to fight off a wild animal, but not the constant stress we experience in the modern world. In the early stages of inflammation, cortisol circulates at higher levels, as the body fails to subdue the inflammatory response, and this can lead to cortisol resistance, meaning your cells don't use the cortisol you make. The inflammation by default is allowed to progress, and eventually cortisol levels plummet. The inflammation continues to climb, and you find yourself in hormonal and immune chaos. There have been multiple studies linking stress, HPA axis dysregulation (adrenal fatigue), and autoimmune disease. Rheumatoid arthritis, which primarily affects the joints, multiple sclerosis, autoimmune thyroid disease, and autoimmune diabetes are but a handful of the conditions linked to stress.

Adrenal Fatigue and the Pill

With HPA axis dysregulation, there's a hiccup in how the brain and the adrenals are talking, and you can also have cellular resistance. This can manifest as what is typically called "adrenal fatigue," which is more accurately termed HPA dysregulation. Your adrenals aren't designed to give out like your ovaries do in menopause and don't really become fatigued. I understand "adrenal fatigue" is a common term, but I want you to understand that it isn't an accurate description of what is really going on. When the pill interferes with the OAT axis, it affects your adrenals, thus causing the HPA dysregulation that results in symptoms associated with "adrenal fatigue."

Because that bolus of estrogen in the birth control pill is highly inflammatory, it causes your adrenal glands to release cortisol in an attempt to squash that inflammation. The brain tells the adrenal glands, "We've got inflammation! Please send cortisol!" The problem is that the daily dose of the pill can be too much for the adrenal glands to handle over a long period of time. To make matters worse, the pill raises levels of a protein called cortisol-binding globulin, which grabs onto cortisol and keeps the body from being able to use it. So the adrenals push out cortisol as they've been instructed, but the cortisol-binding globulin grabs onto it so it can't do its work, and inflammation continues to rise as you continue taking the pill. Over time, the brain and the adrenal glands stop talking in a way that's effective for the body.

If this were a one-off incident, it would be fine, but the consistent estrogen and triggering of inflammation can be just too much for the body's HPA axis. Plus, the birth control pill depletes nutrients important to the health of the adrenals, thereby contributing to adrenal fatigue.

How do you know if you have HPA dysregulation? Take the following quiz to see if you have any of the common symptoms.

Quiz: Do I Have HPA Axis Dysregulation?

Check any boxes that apply to you in the following list:

- ☐ I have difficulty waking up in the morning.
- ☐ I experience fatigue or have low energy.
- ☐ I feel "wired and tired" at night.
- ☐ I crave sugar, salt, and/or carbs.
- ☐ I'm frequently ill.
- ☐ I have hormone imbalance, significant PMS, and/or menopausal symptoms.
- ☐ I have acne and other skin problems.
- ☐ I feel depressed and/or irritable.
- ☐ I have a low libido.
- ☐ My memory is poor.
- ☐ I feel dizzy, light-headed, or a "head rush" when rising from a lying or seated position.
- ☐ I am unable to cope with stress.
- ☐ I have low blood pressure.
- ☐ I feel anxious.
- ☐ I have weight gain around my midsection.
- ☐ Any wounds I get heal poorly.
- ☐ My skin is darkening.
- ☐ I've had unexpected weight loss.

TOTAL _____

1 box checked = low risk of HPA axis dysregulation

2 boxes checked = intermediate risk of HPA axis dysregulation

3 or more boxes checked = high risk of HPA axis dysregulation

Lab Tests for HPA Axis Dysregulation

If you're experiencing any of the symptoms listed in the quiz, then I recommend having the following tests:

- **Adrenocorticotropic hormone (ACTH) with cortisol:** Typically this is tested around 8 a.m., and the results can help you understand how your brain and your adrenals are talking.

- **DHEA-S:** This steroid hormone is produced in the adrenal glands and may be high in the early stages of HPA dysregulation but may drop as your body produces more cortisol at the expense of your other sex hormones. DHEA ultimately gets converted to estrogen and testosterone.

- **Salivary or urinary 4-point cortisol:** This involves testing your cortisol at four different times during the day since levels can vary. This can provide a more detailed picture. Testing urinary cortisol using a DUTCH adrenal test has the benefit of showing inactive cortisol known as cortisone.

- **21-hydroxylase antibodies:** If your symptoms are pronounced, then looking into autoimmune disease is warranted.

Testing for HPA dysregulation can be tricky, which is why it's helpful to look at a variety of tests paired with your symptoms so you receive an accurate diagnosis and the right treatment for you.

It's Not "All in Your Head"

By the time Cali came to my office, she had seen ten other doctors, and her last physician had recommended that she see a psychiatrist for her stress. Cali *was* stressed, but her symptoms were also very real. She suffered from fatigue, anxiety, and daily joint pain, not to mention constipation. She also had been on the birth control pill for about ten years and wondered about the impact it was having on her health as she struggled between having enough energy and feeling super stressed out on a regular basis.

I recognized that Cali indeed needed help with her stress, so I asked her what she did to handle it. I think it's important to recognize and disrupt stress patterns in the moment. It may not be easy, but it can rewire your brain. Cali admitted that when she felt stressed, she ate all the things. Especially salty things. No judgment. We've all been there. But because Cali had felt judged by her other doctors, she hadn't mentioned this habit to them.

Cali's answer to the stress question gave me two important clues into her situation. One, Cali ate when stressed. When you're in a state of stress, blood is directed away from your gut and to your large muscles. Because of this, Cali's gut wasn't working properly and she couldn't digest her food. No wonder she was constipated! Plus, the constant stream of estrogen from the birth control pill wasn't helping. Two, Cali specifically craved salty foods, which was a clue that her adrenals were involved. In addition to cortisol, the adrenal glands also regulate your blood pressure—and salt has a thing or two to do with that system. This is why you crave salt when you're stressed.

Cali needed adaptogenic herbs (more on these on page 145) to get those adrenal hormones balanced plus herbs to reduce her stress response. I prescribed a combination of Rhodiola, ashwagandha, passionflower, ginseng, and licorice root. We also fed her adrenals with B vitamins, vitamin C, healthy fats, and quality proteins. And Cali decided that it was time to quit the pill and revitalize her health.

Unfortunately, none of Cali's prior doctors had gotten to the bottom of *why* Cali had inflammation in the first place. After a few tests, we discovered that Cali had a number of issues, including elevated inflammatory proteins and leaky gut, and she was hypothyroid with elevated antibodies revealing Hashimoto's. I started her on a low dose of natural desiccated thyroid hormone, because with her low T4 and T3, she was having trouble making and converting thyroid hormone. Finally, her poor gut function meant that she wasn't able to absorb the nutrients her thyroid needed, and therefore her gut and thyroid were getting worse and worse.

After two weeks on this regimen and off the birth control pill, Cali's digestion improved, she had more energy, and she experienced less anxiety. And instead of eating every time she felt stressed, Cali learned to take a pause and ask herself if food was really what her body needed in that moment. I also realized that Cali had spent too many years on a restrictive diet that had caused nutrient depletions, as had the birth control pill, so my priority was for her to reintroduce foods and figure out which ones were actually problematic so she could return to eating a more varied diet. She needed healing foods and herbs to get her gut back on track and restore her microbiome. Within six months, Cali's constipation disappeared, her thyroid antibodies decreased, she no longer had anxiety, and, best of all, her fatigue had been replaced with boundless energy.

The Brighten Thyroid and Adrenal Health Protocol

In my practice, many of my patients report that while a thyroid medication does help them, it isn't enough without diet and lifestyle strategies to significantly improve their condition. However you choose to tackle your thyroid disease, you can incorporate appropriate supplements and nutrition into your diet to support thyroid function and facilitate activation of thyroid hormone. Because the gut and thyroid have such a strong influence on each other, I recommend starting with the Brighten Gut Repair Protocol in chapter 6. You'll also want to complete the liver detox in the 30-Day Brighten Program (see chapter 12). Given how the pill negatively affects thyroid health, I would seriously consider quitting it if that is an option for you.

The key steps to the Brighten Thyroid and Adrenal Health Protocol are **ALTER:**

> **A**drenal nourishment and adaptogens
>
> **L**iver detox
>
> **T**hyroid-boosting nutrition
>
> **E**xercise and stress reduction
>
> **R**oot cause of dysfunction

Iodine and Autoimmune

Be careful not to take high-dose iodine supplements. High levels of iodine are used to suppress the thyroid and have been shown to aggravate Hashimoto's thyroiditis. Instead, opt for a combination of selenium and iodine, and limit iodine to no more than 300 micrograms daily from food and supplements.

Adrenal Nourishment and Adaptogens

You will improve the communication of the entire OAT system when you support your adrenals, as well as create balance within your hormone system. Adequate, restful sleep is absolutely vital to restoring your adrenals, thyroid, and hormones, and repairing your body. In addition to sleep, build in time to decompress during your waking hours. Schedule time to relax, breathe deeply, and feel gratitude—this will signal your body to rest as well.

Your adrenals are the foundation of your hormones and you need them to be strong in order to support your other hormones. Be sure to feed them well with vitamins, minerals, and herbs. I recommend taking the following:

- a quality B complex vitamin
- 1,000 to 4,000 milligrams of vitamin C
- 150 to 600 milligrams of magnesium at night

When we're stressed, we actually excrete more magnesium, and most of us are magnesium deficient already because of the deficiency in our food supply. If you're struggling with any kind of hormonal symptoms, I recommend that you have magnesium as a supplement of at least 300 milligrams a day.

One of the best ways to get a head start on supporting your adrenals is with adaptogenic herbs. Adaptogenic herbs enhance adrenal function and immunity as well as physical and mental endurance. They improve the communication between your brain and your adrenals and balance your cortisol output. The following are a few adaptogenic herbs to consider:

- **Maca** acts on the adrenal glands to support healthy estrogen and testosterone levels. Incorporate maca into your daily smoothie!

- *Eleutherococcus senticosus*, or **Siberian ginseng**, is known for improving energy, stamina, focus, and concentration. It helps eliminate fatigue, support the immune system, improve memory and mental alertness, and reduce insomnia. It also helps build muscle. Avoid this herb if you have high blood pressure, and be sure to take it before noon to ensure a good night's sleep.

- **Gotu Kola** can be leveraged if you've been under constant stress or stress is about to ramp up. It will keep your adrenals healthy while you work on decreasing whatever is overstimulating them.

- **Rhodiola** helps balance your cortisol output, reduce the effects of stress, and improve resiliency—think of it as an endurance herb. It improves your energy, lowers anxiety and inflammation, and supports your immune system. Added bonus: Rhodiola also supports healthy progesterone. (Don't take it without supervision if you have a history of depression or bipolar disorder.)

- **Ashwagandha** puts the "hush" on cortisol and brings a deep sense of calm. If you feel "wired and tired," this herb can be especially helpful at bedtime. It reduces inflammation, oxidative stress, and anxiety, and improves sleep, memory, energy, and libido. Research has shown that it also may reduce heavy bleeding during your period and eliminate uterine fibroids with long-term use. Ashwagandha is a nightshade, so it may be problematic if you have a sensitivity to the nightshade family.

- *Cordyceps sinensis* is a medicinal mushroom that helps nourish the adrenals and support their overall function as well as improve liver function. Added bonus: it is traditionally known as a natural aphrodisiac. So skip the champagne this Valentine's Day and heat up your relationship with some *Cordyceps*.

- **Licorice root** increases energy and decreases inflammation by keeping cortisol around longer. Do not take this herb if you have high blood pressure.

- **Reishi** is a mushroom that builds your adrenal strength in addition to keeping colds at bay and making you more resilient to stress.

While these herbs can be tremendously helpful, they don't work overnight, so don't expect results within days. Give them two to three months to really work their magic.

Liver Detox

In chapter 5, you learned about the importance of detox and liver health. Supporting your liver is also essential to thyroid health because environmental toxins and medications can interfere with thyroid function and ultimately cause thyroid disease. The liver too is a major site of thyroid conversion, helping that T4 hormone to turn into active T3. Your liver is also a key organ in blood sugar (glucose) regulation, which is intimately tied to your adrenal glands. Cortisol causes the liver to break down glycogen (the storage form of glucose) and release glucose into the bloodstream in response to stress, which includes skipping meals. Studies have shown that adrenal and liver dysfunction accompany each other. And remember all that cholesterol-is-used-to-make-hormones talk back in chapter 2? Well, your liver is responsible for making the cholesterol used to synthesize your hormones. What doesn't the liver do, right? If you haven't already, begin the liver detox in the 30-Day Brighten Program in chapter 12.

Thyroid-Boosting Nutrition

First and foremost, ingest foods and supplements that will optimize your thyroid health. You can begin to replenish your missing nutrients by incorporating grass-fed meats, seafood, leafy greens, mushrooms, and other vitamin- and mineral-packed foods into your diet.

The following nutrients are the most important for thyroid health:

- **Selenium** is needed in the production of T4 and its conversion to T3. A daily dose of 200 micrograms has been shown to be beneficial to thyroid hormone production and to reverse and prevent autoimmune thyroid disease. Add organ meats, seafood, Brazil nuts, and muscle meats to your diet.

- **Iodine**, along with the amino acid tyrosine, is what thyroid hormone is composed of, so iodine is pretty important for overall thyroid health. You need only 150 micrograms daily, which you can get by consuming seafood and seaweeds. Iodized table salt and other foods contain iodine, so be careful not to overdo it or it can have a detrimental effect on thyroid synthesis.

- **Omega-3 fatty acids** keep cell membranes strong, allow thyroid hormone to enter those cells, lower inflammation, and help maintain thyroid gland function. Essential fatty acids also help balance your hormones and stabilize your energy. You can find essential fatty acids in herring, salmon, sardines, oysters, and other cold-water fatty fish.

- **Magnesium** helps regulate blood pressure, which can spike when you have a thyroid hormone deficiency. Increase your magnesium intake by eating plenty of green vegetables, lima beans, fish, and nuts.

- **Zinc** does a lot when it comes to your thyroid. Along with other vitamins and minerals, it helps with the synthesis of thyroid hormone and the conversion of T4 to T3. It also helps your brain (the hypothalamus specifically) detect and respond adequately to the thyroid hormone circulating in your bloodstream. Zinc can be found in shellfish, especially oysters, and red meat. Unfortunately, the zinc you might find in whole grains, legumes, and nuts is not bioavailable because it is bound to phytic acid.

- **Iron** plays a crucial role in the production of thyroid hormone, and a deficiency is common in people with hypothyroid disease. It assists with the conversions of T4 to T3 and iodide to iodine, which lead to you feeling energetic, having easier periods, and feeling in love with life. You can find iron in animal sources, like chicken, beef, pork, and seafood, as well as dark leafy green vegetables. Interestingly, an iron deficiency usually accompanies a vitamin A deficiency. Studies have shown supplementation of vitamin A to have beneficial effects on iron status. However, note that vitamin A is considered a teratogen, which means if you're supplementing with a high dose, it could be harmful if you become pregnant. Women who have periods should aim to get 18 to 30 milligrams daily of iron, but your dosage may vary based on your individual needs.

- **Vitamin A**, as an antioxidant, assists your immune function. It improves cellular sensitivity to thyroid hormone. When vitamin A is ready for use in the body, it is called retinol, and you can find it in animal products like liver, egg yolks, grass-fed butter, and cod liver oil. Carotenoids can also be converted to vitamin A when your body needs it, and you can find them in sweet potatoes, carrots, mangos, apricots, orange-colored vegetables, and leafy green vegetables. If you're actively trying to become pregnant or not using any form of birth control, aim for only beta-carotene sources of vitamin A in your supplements and eat plenty of carotenoid-rich vegetables, as this the safest form of vitamin A during pregnancy.

- **B vitamins** are involved in thyroid hormone production, immune function, and detox, which are all areas in which women on the pill (and women in general) need support. If you've been on the pill or are currently taking it, a B complex can help replenish those nutrients the pill depletes while also supporting hormone balance. The Brighten Protocol diet includes food sources of B vitamins.

- **Vitamin D** is super important for thyroid health because it is involved in transporting thyroid hormone into cells and it supports your immune cells. One of the easiest ways to get vitamin D is to spend time in the sun, but you can also find it in salmon, sardines, mackerel, or fish liver oils. Getting tested is the best way to determine if you need a supplement and the dose that will work best for you. In general, most people benefit from a 1,000 to 2,000 IU supplement daily.

Follow the 30-Day Brighten Program meal plans in chapter 12 to energize your thyroid hormones and adrenal health.

WHY YOU SHOULD QUIT GLUTEN IF YOU'RE HYPOTHYROID

In chapter 6, I discussed why it's necessary to eliminate gluten if you're having gut issues, but it's especially important if you are hypothyroid, because 90 percent of hypothyroid cases are caused by an autoimmune condition. In cases like Hashimoto's thyroiditis, antibodies signal the attack on your thyroid, destroying the thyroid tissue. There is mounting evidence that gluten drives this attack on your thyroid. If you have a food sensitivity or leaky gut (remember, the pill causes this), when a protein in gluten, called gliadin, enters your bloodstream through the gut, your immune system sees the gliadin as non-self and releases antibodies. The problem is that your immune system mistakes your thyroid for gliadin in a process known as molecular mimicry. Essentially, every time you eat gluten, your body produces antibodies to gliadin, which then ultimately trigger an attack on your thyroid.

Many of my patients have lowered their antibody levels and reduced their uncomfortable symptoms through dietary practices like eliminating gluten, and you can too.

Exercise and Stress Reduction

As with all the healing protocols in this program, it's important to get your body moving regularly, because it lowers inflammation and improves hormones. When it comes to the thyroid, exercise also supports the conversion of T4 to active T3. If thyroid disease or adrenal symptoms are making you feel fatigued, start out small and increase your time and intensity as you're able. Aim for thirty minutes of movement daily—this can include walking, weight training, cardio, yoga, or whatever you enjoy. Movement enables you to release stress and tension from your body, increase oxygen to your brain, and counter the effects of cortisol.

While incorporating exercise into your lifestyle as a habit will have a positive effect on your thyroid—and overall—health, just be careful not to overexercise or engage in high-intensity activities for too long, such as running for two hours or taking a rigorous spin class or boot camp workout. This can add too much stress, which as you already know can lead to HPA dysregulation and the inflammation that derails thyroid hormone production. Stress has also been shown to alter gut bacteria, which can have a detrimental effect on your thyroid and adrenals. In addition to light exercise, try deep breathing, meditation, and mindfulness to alleviate stress when your life feels out of control. Even if stress is virtually impossible to avoid in today's world, you *can* choose how you respond to it. Try one of the stress-reducing practices listed on page 260. Kick your stress to the curb and discover your inner Zen.

Root Cause of Dysfunction

If you have any infections, parasites, or critters growing in your gut, you will need to treat them in order to lower your inflammation, restore your gut, and heal your thyroid and adrenals. Many natural therapies can play a part in rebalancing your gut. See the Brighten Gut Repair Protocol in chapter 6.

Key Takeaways:
Energize Your Thyroid and Adrenals

- When prompted by TSH, your thyroid secretes inactive T4 and a small amount of T3. The T4 gets converted to active T3 in the gut, liver, kidneys, and other tissues.

- The two types of thyroid disease are hyperthyroidism (too much thyroid hormone) and hypothyroidism (not enough thyroid hormone), with the latter being most common.

- The primary symptoms of hypothyroidism are weight gain, cold intolerance, depression or anxiety, slow heart rate, delayed reflexes, fatigue, memory loss, thinning hair, dry skin, brittle nails, and constipation.

- In addition to TSH, your doctor should test total T4, total T3, free T4, free T3, reverse T3, anti-thyroperoxidase (anti-TPO), and anti-thyroglobulin antibodies.

- The pill depletes important nutrients that your thyroid needs and also increases thyroid-binding globulin (TBG), binding up any thyroid hormone you do manage to make.

- Leaky gut increases the risk of thyroid disease, inhibiting active thyroid production. Remember, 20 percent of thyroid hormone conversion occurs in the gut.

- The pill also causes HPA axis dysregulation (adrenal fatigue) because the constant stream of estrogen produces inflammation, causing the adrenals to push out more cortisol.

- The Brighten Thyroid and Adrenal Health Protocol includes the following steps: consume nutrients that support thyroid function, begin moderate exercise and decrease stress, support your liver, find the root cause of infections or inflammation, and support your adrenals.

- The most important nutrients for thyroid health are selenium, iodine, essential fatty acids, magnesium, zinc, iron, and vitamins A, B, and D.

- You can begin to support your adrenals with the following adaptogenic herbs: maca, *Eleutherococcus senticosus*, Gotu Kola, Rhodiola, ashwagandha, *Cordyceps*, licorice root, and reishi mushroom.

CHAPTER 8

REVERSE METABOLIC MAYHEM

Metabolism is about much more than the number on the scale. Your metabolism causes the chemical reactions in your cells that create energy from the foods you consume, and it can play a role in regulating everything from blood sugar level to hormonal balance. And when the birth control pill interferes with proper metabolic function, it can set off serious issues that I call metabolic mayhem.

The pill causes inflammation and insulin resistance, which can lead to blood sugar imbalance, which puts you at risk for three life-threatening conditions: heart disease, stroke, and cancer. The pill also elevates your cholesterol and blood pressure, two more risk factors for these serious conditions. Finally, the pill increases your risk of blood clots, which can cause a stroke, especially if you have certain genes.

In this chapter, I will help you determine if you could be at risk for stroke, heart disease, or cancer so you can make an informed decision about whether the pill is right for you—especially if you're taking it for non-contraceptive reasons. Whether or not you choose to remain on the pill, the Brighten Metabolic Protocol can support you with some lifestyle changes that will tell metabolic mayhem exactly where it can go.

Blood Sugar Imbalances and Your Hormones

Have you ever had one of those high-stress days when you didn't have time to eat and grabbed a quick muffin during an afternoon meeting only to have your hangry self seemingly come out of nowhere? And then you wanted to face-plant on your desk with exhaustion afterward? Yeah, well, you can blame it on your blood sugar. Fluctuations in your blood sugar can give your hormones some legit whiplash and result in symptoms of irritability, fatigue, and even nausea. While a healthy diet can help you balance your blood sugar—and hormones— it can help only so much if you're struggling with the inflammation caused by the pill. Inflammation can lead to insulin resistance and high blood sugar, also known as the road to metabolic mayhem.

A 2016 study published in *Endocrine* revealed that past use of the pill for more than six months significantly increased the risk of developing diabetes in postmenopausal women. Nondiabetic pill users were found to have significantly elevated levels of insulin, which is a sign of insulin resistance and a risk factor for developing diabetes and heart disease. Pill users were also diagnosed with diabetes at a younger age. Researchers found, too, that women who had used the pill had higher cholesterol and blood pressure compared to those who had not taken the pill for an extended period of time. What research suggests is that using the pill for over six months during reproductive years is a significant risk factor for developing diabetes. Your doctor probably forgot to mention this (insert eye roll).

The Pill, Insulin Resistance, and PCOS

Okay, so the pill leads to insulin resistance and elevated blood sugar in some women. You know what else causes insulin resistance? Polycystic ovary syndrome (PCOS). PCOS is one of the primary non-contraceptive reasons women are prescribed the pill. Many doctors prescribe it to help with the symptoms of acne, weight gain, excessive hair growth, and irregular periods.

In my opinion, giving a PCOS-diagnosed woman the pill to mask her hormonal symptoms without at least covering the side effects of the pill is not only a disservice but downright dangerous. (Hello, heart disease and stroke!) As you now know, the pill is a hormonal Band-Aid and doesn't actually get to the root cause of your symptoms, which means these conditions are left to progress

In This Chapter

- How the pill can make you feel bitchy and exhausted (Hint: check your blood sugar.)

- What's the deal with cancer risk

- Why that annoying chin hair and acne might not be PCOS

- How remaining on the pill indefinitely can be downright dangerous

- Why it's a bad idea to take the pill if you get headaches or migraines

- Why you should get tested for human papillomavirus (HPV)—and quit the pill if you have it

- How the pill can and does kill

while you depend on your daily pill. And evidence suggests that insulin resistance resulting from inflammation may actually *cause* PCOS, not the other way around.

Remember from chapter 2 that insulin is a hormone produced by the pancreas that helps bring blood sugar into your cells. Inflammation causes your cells to become rigid, and as such, they're no longer able to receive the signal from insulin to transport sugar into your cells. When all that sugar is left to wander your bloodstream, your pancreas makes more insulin. And more. And more. Until your cells get so tired of insulin always knocking on the door that they stop listening and ignore it, which ultimately leads to insulin resistance. This causes some pretty serious damage to your brain, kidneys, eyes, and ovaries.

Your ovaries do *not* become insulin resistant, but if they are bombarded with insulin, a structural change occurs and stimulates your ovaries to secrete testosterone. This causes hair growth on your chin, chest, and abdomen. WTF! Insulin also inhibits the secretion of sex hormone–binding globulin (SHBG), a protein made in the liver that binds to sex hormones to prevent any symptoms that can come from too much testosterone or estrogen. So, while on the surface PCOS may seem like a sex-hormone-only issue, it has roots in metabolic disorder. Blood sugar dysregulation is at the crux of many hormone imbalances, including elevated testosterone, and it can lead to infertility. But here's the tricky part: some women coming off the pill experience the same symptoms of PCOS.

POST-PILL PCOS

You can experience symptoms of PCOS without actually *having* PCOS due to insulin and blood sugar issues driving testosterone levels up. (Your blood sugar is seriously the boss when it comes to your hormones!) This is commonly called post-pill PCOS, or pill-induced PCOS, as first defined by Dr. Lara Briden, author of *Period Repair Manual*, and it is one of the ways post–birth control syndrome can also manifest. But it's not *really* PCOS.

Women mistakenly get diagnosed with PCOS because the pill leads to insulin resistance, suppresses ovulation, and can cause testosterone to climb once you stop taking it. A testosterone increase can be due to blood sugar issues or what is referred to as the androgen rebound, which is the increase in testosterone production that can occur when you stop the pill. I've met with countless patients who ditched the pill only to find their skin flare up with relentless and painful acne. You see, when you stop the pill, if your ovaries kick up testosterone production, the oil glands in your skin kick out sebum. This is part of why post–birth control acne can be so bad and why taking the pill to try to clear it up is a bad idea long term.

How do you know if your symptoms are not really PCOS? Women with pill-induced PCOS have a history of regular periods and had no signs of insulin resistance before starting the pill. What I've seen clinically is most women with post-pill PCOS do not have a period, have elevated androgens (i.e., testosterone), and often don't have blood sugar/insulin issues to the extent a woman with true PCOS does. Occasionally the insulin issues are there, and then I look at the woman's menstrual history. Classic PCOS women have *never* had a regular cycle.

When Olivia came to see me, she was in her early thirties and had gone off the pill in hopes of having a baby, only to find that she no longer had a period. Her doctor had diagnosed her with PCOS, but it didn't add up with her history of regular periods *before the pill*. Her menstrual cycles always had been about twenty-seven to twenty-nine days every cycle and no other symptoms from her youth painted a PCOS picture. But when Olivia came off the pill, she did not get a period. What did she get instead? Oily skin, tons of acne, and hair loss, which she was really sensitive about. By the time I saw her, Olivia hadn't had a period in eight months! Based on slight blood abnormalities and her symptom picture, her doctor had delivered the PCOS diagnosis and told her she would have to undergo fertility treatments if she wanted to get pregnant. A doctor herself, she was confused by this diagnosis, so she sought me out to help her discover why she was having these symptoms.

Olivia's blood work revealed that she did not have insulin resistance, as evidenced by her insulin levels, hemoglobin A1C, and fasting glucose. She had a high-sensitivity C-reactive protein (hs-CRP) measure of 3 mg/L, meaning she had elevated inflammation in her body, which pointed to an adrenal–ovary connection. We also examined her follicle-stimulating hormone (FSH) and luteinizing hormone (LH) levels. She did have a slightly higher level of LH, but not as high as typically seen with PCOS. And her ultrasound showed no cysts on her ovaries.

Like a typical doctor, Olivia was working a lot of hours and not eating lunch. She had symptoms of hypoglycemia and clear signs she was making a lot of testosterone, which led to the oily skin and acne. She recalled, "About three months after I stopped the pill, I started getting hard painful bumps along my jaw, and I was constantly blotting my face because of all the oil." My first goal was to help Olivia get her acne under control by eating better, so I had her start the 30-Day Brighten Program to help rebalance her blood sugar. I also recommended circadian rhythm support and brought in herbal and nutritional therapy, like saw palmetto, turmeric, omega-3s, zinc, nettle root, and freshly ground flaxseeds to begin reducing the oily skin and acne. In addition, I got her going with laser light therapy for her scalp to stimulate the growth of her hair.

Because it had been quite a while since Olivia had had a period, we began syncing her cycle to the moon with seed cycling (see page 258) and herbal tinctures. From new moon to full moon, she took higher doses of black cohosh. From full moon to new moon, she took Vitex and used micronized progesterone cream topically, because it has an antiandrogenic effect that helps block the testosterone.

Olivia was able to have a normal period within only three months and transition off the progesterone, but I've seen it take eighteen to twenty-four months and sometimes longer to restore a period, so don't lose hope. She continued using the herbs a bit longer, and she never did go on to develop insulin resistance.

While on the 30-Day Brighten Program, Olivia realized that dairy was a trigger for her. She had no problem steering clear of dairy if it meant the days of acne were behind her. Happily, seven months after Olivia started the program, she was able to conceive naturally. And her period came back postpartum at a regular twenty-nine-day cycle.

The Pill and Metabolic Syndrome

Metabolic syndrome is a group of conditions (three or more) that can increase your risk of heart disease, stroke, and diabetes. They include abdominal obesity, high blood sugar, high blood triglycerides, low HDL cholesterol, and high blood pressure. You now know that the pill can contribute to weight gain and high blood sugar, because of inflammation and insulin resistance, but it also affects your cholesterol and blood pressure. If you're at risk of or have a family history of metabolic syndrome, you may want to consider one of the alternative forms of contraception in chapter 13.

The pill has been shown to lead to elevated total cholesterol, high LDL (bad cholesterol), low HDL (good cholesterol), and elevated triglycerides in some women. PCOS also has been linked to higher triglycerides, and because oral contraceptives can cause those levels to rise further, taking the pill when you have PCOS may increase the risk of cardiovascular disease. If your doctor's only solution to your PCOS symptoms is to put you on the pill, find another one! For real. We have better ways to clear up that hair growth and acne without putting your life at risk.

Vitex and PCOS

The use of Vitex in women with PCOS has been controversial because it was always thought to raise luteinizing hormone, and women with PCOS already have too much LH. But Dr. Fiona McCulloch, author of *8 Steps to Reverse Your PCOS*, argues that women with PCOS and PCOS-like symptoms can develop a dysfunction in the opioid system of their brains and Vitex can help reset that system. Insulin resistance causes that dysfunction. So Vitex can be beneficial to women with PCOS, contrary to popular belief.

Elevated Blood Pressure

Hormonal birth control can also increase your blood pressure, and this is often listed as a side effect right on the packaging, with the risk increasing if you are older than thirty-five, overweight, or a smoker. Over long periods of time, high blood pressure can lead to damage of the kidneys, eyes, and brain. In fact, high blood pressure is a risk factor for a type of stroke known as transient ischemic attack. Monitoring your blood pressure regularly is an important screening tool.

Some doctors advocate for putting women on the pill after age thirty-five, claiming that it will manage perimenopause symptoms and that women can just stay on it until they stop having periods (remember, a period on the pill is a withdrawal bleed). It's one thing if you're concerned about pregnancy prevention, but considering the side effects (ya know, like the kind that kill), there should be a discussion about the risks. Also, there are other, less high-risk ways to prevent babies (see chapter 13).

Blood Clots

The pill is known for increasing the risk of blood clots, which result in pulmonary embolism, deep vein thrombosis, and stroke. You're at risk for blood clots if you're overweight, over thirty-five, or a smoker. While many critics have argued

Metabolic Issues and B6 Deficiencies

As we've talked about in previous chapters, the pill depletes many nutrients, including vitamin B6. Studies have found that as high as 75 percent of women taking oral contraception have a B6 deficiency, which is independently associated with an increased risk for stroke and heart disease. A B6 deficiency also contributes to elevations of a molecule known as xanthurenate, which has a negative effect on insulin and has been shown in animal studies to lead to diabetes. If you're on the pill and you need to replenish some B6, foods like avocados, pistachios, sesame and sunflower seeds, chicken, and grass-fed beef are excellent sources.

that the risk is mild compared to that during pregnancy or postpartum, both of these states occupy a finite window of time in a woman's life. The pill is typically used for decades. Ever seen a woman pregnant for decades? Nope. The pill carries a long-term risk for developing a blood clot, which can lead to death. Women of all ages are at risk, but especially as we enter into menopause. If you're between the ages of twenty and forty-four, then it's important you know that 1 in 2,000 women in this age range and on the pill will be hospitalized for abnormal clotting. Compare that to non-users in the same age group and about 1 in 20,000 (yes, that's an extra zero there) will be hospitalized for the same conditions.

The pill can and does kill. In women aged fifteen to thirty-four, the risk of death due to circulatory disorder is 1 in 12,000. In non-users, that risk is about 1 in 50,000. While your doctor may have told you the risk of stroke and death are low, it is important to examine these numbers in the context of your personal and family history. Also, maybe 1 in 12,000 seems low, but when it's you or your sister or your best friend, that number becomes a lot more significant. And it leads many women to ask, "Just how necessary is this pill?"

Women on combined estrogen-progestin pills also have a two- to four-fold increased risk of a venous thromboembolism (VTE) compared to women not on the pill. A VTE is when a clot develops within your vein. You may be familiar with the term deep vein thrombosis (DVT), which is a clot in a deep vein, usually in the leg but sometimes in the arm or other veins. DVTs become especially dangerous when they break free from the vein wall, travel to the lungs, and block some or all of the blood supply. This is known as a pulmonary embolism (PE). DVTs and PEs can place you at greater risk of a stroke. While manufacturers have tried to lower the VTE risk in newer versions of the pill, it still persists. The lowest risk seems to be with levonorgestrel-containing contraceptives, although if you have a family risk of blood clots or stroke, or are at an increased risk of a clot due to genetic or other factors, then I advise you to reconsider the use of hormonal contraceptives.

This risk of developing a clot is a real concern. So much so that in 2012, based upon available data, the US Food and Drug Administration added revised labeling to all oral contraceptives containing drospirenone, stating that they may be associated with up to a three-fold higher risk of VTE. Drospirenone is a synthetic progestin used in combination birth control pills, like Yasmin, and it has been linked to a higher risk of blood clots and stroke.

If you're on the pill to treat acne, PMS, headaches, or other hormonal symptoms, I urge you to find the root cause of your symptoms. Using the pill to treat symptoms when there are safe and effective alternatives may be putting your life at a significant and unnecessary risk.

The Pill and Strokes

The risk of stroke is well documented but quickly dismissed by many doctors because it's not considered significant. Um, excuse me? While the increased risk may be small, it is *not* insignificant, and individualized counseling is in order. The ischemic type of stroke associated with the pill is caused by a blood clot obstructing flow of oxygenated blood to the brain. In general, about 4.4 ischemic strokes occur for every 100,000 women of reproductive age. The pill almost doubles the risk to 8.5 strokes per 100,000 women, according to a meta-analysis.

Women with lupus anti-phospholipid antibodies are one of the highest risk groups for stroke and heart attack. The odds of these women having a heart attack while on the pill increased to 21.6 per 100,000.

Women with elevated levels of the amino acid homocysteine while on the pill are also at higher risk of a blood clot in the brain. I don't know about you, but a blood clot in the brain? No, thank you. I recommend some really simple blood tests and a screening (see page 164) if you're on the pill or considering starting hormonal birth control. This is the best way to understand your individual risk.

What most women aren't told about the stroke risk is that smoking cigarettes, high blood pressure, obesity, certain genes, and a history of migraines with auras (flashes of light) increase the risk significantly. Women with these habits or medical conditions should *not* use the pill. According to the CDC, the following items pose a health risk for women using a combined estrogen-progestin oral contraceptive:

systemic lupus erythematosus (positive or unknown anti-phospholipid antibodies)

personal or family history of blood clots or stroke

age thirty-five or older

smoker of fifteen or more cigarettes a day (let's be clear: that doesn't mean it's okay to smoke a few or even one daily!)

excess body weight

migraines with auras (doesn't matter how old you are)

high blood pressure (systolic ≥160 mmHg or diastolic ≥100 mmHg)

high cholesterol

breast cancer

liver disease, including cirrhosis, tumors, or cancer

heart or lung disease, including atrial fibrillation or pulmonary hypertension

having multiple risk factors for heart disease, like advanced age, smoking, diabetes, and high blood pressure

having a genetic mutation that increases the risk of clotting, such as factor V Leiden and/or *MTHFR* gene variations

Genetic Risk and the Pill

If you're taking the pill, it's important to know whether you have factor V Leiden and *MTHFR* gene variations, and you can easily get tested. In my clinic, I screen every woman for these variations so I can provide individualized recommendations. The Risk of Arterial Thrombosis in Relation to Oral Contraceptives (RATIO) study found that women who had a single copy of a factor V Leiden gene mutation were at an eleven-fold risk for an ischemic stroke compared with women not using oral contraception.

Additional risk factors for experiencing an ischemic stroke include age. Yes, the older you are, the higher your risk. The estimated incidences of having a clot is 100 cases per 100,000 people for women who are over thirty-nine and on the pill. To give a comparison, for adolescents this number is 25 per 100,000 people.

Another study placed the threat of stroke in **women with the factor V gene taking oral contraceptives as high as a thirty-five-fold increase**. Women with *MTHFR* 677TT using oral contraceptives had a 5.4-fold increased risk of stroke. In my practice, we test for and commonly find *MTHFR* variations. While factor V is a less frequent finding, it's something we screen for. I recommend getting screened for both of these genes before you start using hormonal birth control, and if you're already on the pill, have this checked.

What Are the Signs of a Stroke?

Listen, when I was twenty I didn't know a thing about what it would look like to have a stroke, but that didn't stop me from popping a pill that increased my risk of having one. Some of the typical warning signs of a stroke include:

- numbness or weakness, especially on one side of your body, such as your face, arm, or leg
- feeling of pins and needles
- confusion or trouble with understanding people
- difficulty speaking or slurred speech
- blurred vision, double vision, or loss of vision
- problems with balance or coordination, difficulty with walking
- dizziness, light-headedness, or vertigo
- sudden severe headache

If you experience any of these symptoms, especially more than one at a time, take it seriously and get yourself checked out stat!

Headaches and Migraines

If you or your parents have a history of headaches, then you're at increased risk for developing new headaches when starting the pill. And the new onset of headaches is a big reason why many women ultimately choose to quit the pill. For some women, the headaches can subside after several months on the pill, but for others, they persist. Headaches and migraines are a possible sign of inflammation, hormone imbalance, celiac disease, food sensitivity, autoimmune disease, and more. Their root cause should be investigated.

When it comes to migraines, some women report great improvement after starting the pill, but others report an increase in frequency or duration. (Are you noticing a theme here? It's about what is true for you.) While experiences and the data vary, one thing we do understand well is that women who experience auras with their migraines are at an increased risk of ischemic stroke.

What to Test Before Popping That Pill

I recommend considering the following tests if you are thinking about taking the pill or are already on it, especially if there is a family history of blood clots, stroke, early-age heart attack, or pulmonary embolism:

- anti-cardiolipin antibodies
- anti-phospholipid antibodies
- blood pressure
- cholesterol panel
- complete blood count (CBC)
- comprehensive metabolic panel (CMP)
- factor II genes
- factor V genes
- fibrinogen
- high-sensitivity C-reactive protein (hs-CRP)
- homocysteine
- lipoprotein-A
- lipoprotein-associated phospholipase A2 (Lp-PLA2) activity
- *MTHFR*
- myeloperoxidase (MPO)

The Pill and Heart Attacks

As we previously discussed, the pill also raises the risk of heart attacks, something women have been dying of at higher rates than men since the 1980s. Why do we die at higher rates? Well, that would require an entirely different book, exploring the intricacies of medical gender bias and how it leaves women sick—and sometimes dead! In fact, every 80 seconds a woman in the United States dies from a heart attack. But in short, we present differently, and that sometimes confuses doctors. That said, your primary care physician and ob-gyn are awesome at identifying risk factors for heart disease and other metabolic issues, but they need your data. In addition to any blood work or other tests, they need to know your past medical history, your family history, and about any symptoms you may be experiencing. They can't help you if they don't know all the facts. See the checklist on page 167 for what to talk to your doctor about.

In the United States, Yaz, Yasmin, and Ocella are a few of the more well-known contraceptives that come with a **black-box warning for increased risk of serious cardiovascular events**. These are the same drugs used to treat benign conditions like acne. It seems a little extreme to put a gal on a drug that can cause a stroke or heart attack without giving her a heads-up about the risk, or giving her other, safer options for dealing with acne. In case you're not aware, a black-box warning is a pretty big deal. It is the strictest warning the FDA gives out, and the FDA only gives it when there has been enough evidence that a drug is associated with some seriously bad health consequences.

When Yaz was released in 2006, it quickly became the pill of choice for doctors and women alike. After all, their commercials advertised it as pretty much a cure-all for PMS symptoms with their "Beyond Birth Control" slogan. But, as it turned out, Yaz was never actually approved to treat PMS, and it wasn't more effective at preventing pregnancy than previous pills. And those side effects—well, the FDA didn't appreciate Bayer making claims about PMS or using distracting content in their commercial while the narrator listed off the side effects.

Long before the FDA stepped in, however, women were sharing their stories about the pill all over the internet. Weight gain, worsening PMS symptoms (yes, the very thing the manufacturers touted to treat), headaches, irregular bleeding, anxiety, depression, fatigue, loss of libido, nausea, indigestion, gallbladder disease, heart palpitations, heart attacks, and strokes were being reported. Mind you, these are the same symptoms women have been reporting since the introduction of the pill, but there was something new here, and the women taking the pills were starting to catch on.

What made Yaz so special? It was the first time a contraceptive hormone, drospirenone, was introduced as something to treat symptoms in women and serve a larger role beyond pregnancy prevention. Unlike other synthetic progestins, drospirenone is derived from spironolactone, which gives it the ability to act in your body as an anti-mineralocorticoid (think diuretic—it makes you pee more). It also has antiandrogenic effects (think anti-testosterone), which makes it effective at improving acne. But the diuretic effect that helps with bloating?

That's where things get scary quick. This drug causes you to lose sodium and hold on to potassium by opposing aldosterone, a hormone produced by your adrenal glands to regulate blood pressure. Too much potassium can cause a heart attack, and over time it can lead to kidney damage in certain people. If you already have kidney or liver disease or adrenal insufficiency (as is the case with Addison's disease), or you're taking a medication that causes you to hold on to potassium (like spironolactone or NSAIDs), then you want to reconsider taking this hormone. Commonly used NSAIDs, like ibuprofen and naproxen, also cause you to hold on to potassium, which is why regular blood testing of your potassium is recommended if you're going to use these at a high dose or longer term. Women with painful periods, as is the case with endometriosis, may be at risk if they're relying on NSAIDs for pain relief while taking a hormonal contraceptive that contains drospirenone.

One study assessed the data of 1.6 million Danish women who took a drospirenone contraceptive pill for several years. Researchers found that the risk of a heart attack or stroke was higher in these women than in those who used a non-hormonal method of contraception.

What a Heart Attack Looks Like in a Woman

We gals tend to have symptoms that don't look like those of our male counterparts. Use this list to get familiar with what it looks like for a woman to have a heart attack so that if you ever find yourself experiencing these symptoms, you will know to get help immediately:

- cold sweats
- fatigue
- flu-like symptoms
- indigestion
- light-headedness
- nausea
- pain in arms, back, neck, jaw, or stomach
- shortness of breath
- uncomfortable, intermittent squeezing in the center of the chest

Am I at Risk for Heart Attack?

You may be at increased risk of heart attack if you:

- have a parent who has or had heart disease
- are African American, Mexican American, American Indian, or native Hawaiian
- smoke
- have high blood pressure
- have high cholesterol
- have diabetes
- are overweight or obese
- lead a mostly sedentary lifestyle
- eat a poor diet
- regularly experience stress
- drink alcohol
- are over sixty-five (but heart attacks can and do happen to much younger women)

The Pill and Cancer

Breast, cervical, liver, and brain cancer risks all increase with the pill. Many experts have concluded that the longer a body is exposed to synthetic estrogen—like that in the pill—the higher the risk of developing estrogen-related cancers. And this just in: even those newer, low-dose pills pose a risk. A recent study in the *New England Journal of Medicine* followed 1.8 million women, aged fifteen to forty-nine, and found that hormonal contraceptives were associated with an elevated risk of breast cancer. The study also found that progestins (synthetic progesterone) may increase the risk of breast cancer. Women who used the pill had a 20 percent higher risk of breast cancer than women who used non-hormonal forms of contraception. Let me break that down for you: for every 100,000 women who *do not* use hormonal contraceptives, 55 per year will likely receive a breast cancer diagnosis. For every 100,000 women who are on the pill and other forms of hormonal birth control, 68 per year will likely be diagnosed with breast cancer.

More long-term studies are needed, but early evidence suggests that this risk does not disappear once you stop using the pill but instead reveals a long-term elevated risk. The takeaway? All forms of hormonal contraceptives raise the risk of breast cancer. While current studies suggest the risk is modest, if you have a history of breast cancer in your family, you'll want to consider alternative forms of birth control. If you're currently taking the pill or have used it in the past, you can implement the changes in the Brighten Metabolic Protocol on page 171 to start supporting your body in eliminating estrogen and boosting cell-protecting antioxidants.

Now, the pill isn't all bad when it comes to cancer; incidents of uterine, endometrial, ovarian, and colorectal cancers are reduced in women who take the pill. It also appears the protective effect may extend beyond when a woman stops taking the pill—although studies have shown that insulin resistance increases the risk of endometrial cancer, so the debate continues on how well it protects against this form of cancer. I would argue that there are far better ways to prevent cancer that can benefit your entire body and come with much fewer side effects. Let's get real—how helpful is it if the pill reduces some cancers only to increase the risk of others?

A systematic review of twenty-eight studies that looked at more than twelve thousand women with cervical cancer found that cervical cancer risk increases the longer the pill is used. The risk actually doubled after ten years of use! The risk increased even after accounting for factors like the number of sexual partners, the use of barrier contraceptive methods, previous pap smears, smoking, and human papillomavirus (HPV) status.

The Collaborative Group on Epidemiological Studies of Cervical Cancer found in their analysis of twenty-four studies of more than sixteen thousand women with cervical cancer that this **risk of invasive cervical cancer increased the longer a woman stayed on the pill**. The good news is, they also found the risk decreased once the pill was stopped, and after ten years off the pill, the risk was similar to that of women who had never used the pill. Now, the tricky thing about these studies is that they were not able to assess the full impact of HPV, and other studies have found that HPV-negative women on the pill do not demonstrate as high of a risk. That said, about 79 million Americans have HPV, and that's about 80 percent of sexually active men and women.

Here are a few facts to consider about HPV:

- An in vitro study showed that estradiol stimulates growth of HPV-positive cervical cancer cells.

- Younger women who have the highest exposure to HPV are on oral contraceptives—OMG.

- Taking synthetic estrogen increases cervical ectopy (when the inner cells of the cervix are on the outside), and HPV loves to hang out there.

- Taking estrogen also increases estrogen receptor expression—HPV hooks into that receptor.

The Pill, HPV, and Cervical Cancer Risk

The human papillomavirus (HPV) is a sexually transmitted virus with more than 150 strains. About 13 of them are known to cause cancer. Unlike other

---

Can Melatonin Reduce the Risk of Breast Cancer?

Melatonin, often known as the "sleep hormone," is mainly produced by the pineal gland but is also manufactured in a few other places, such as the gut and the white blood cells. Melatonin is a great antioxidant and protective against breast cancer, which is another reason why sleep is so essential. If your melatonin is low and you have potent estrogens (like 4OHE1 or 16OHE1), you could be in big trouble when it comes to breast cancer. Melatonin receptors are found on ovarian cells, indicating that melatonin may play a role in the estrogen levels produced in your ovaries, acting as a selective estrogen receptor modulator and inhibiting its growth. Studies have revealed that night-shift workers, like nurses, are at higher risk for breast cancer, and women who have been diagnosed with breast cancer have lower levels of melatonin. A meta-analysis found a 40 percent increase in breast cancer in women who worked night shifts. Another meta-analysis of flight attendants revealed a 44 percent increased risk of developing breast cancer versus the general population, which is not surprising when you consider that many flight attendants work at night and cross time zones, potentially disturbing their circadian rhythms.

sexually transmitted infections, HPV is common—there's a good chance that you or someone you know has it. About 70 percent of cervical cancer, 95 percent of anal cancer, and 65 percent of vaginal cancer can be attributed to HPV.

Several studies have proposed that what places pill users at the most risk is the metabolism of the estrogen in the pill into the metabolite 16-alpha-hydroxyestrone (16OHE1), which teams up with HPV to cause cancer to grow. Remember from chapter 5 that the liver metabolizes your estrone and estradiol into the metabolites 2OHE1, 4OHE1, and 16OHE1, and that the 16OHE1 pathway can stimulate tissue growth, potentially leading to cancer.

If you are diagnosed with HPV, take a deep breath. It doesn't necessarily mean you will develop cancer. Following your doctor's guidelines for regular monitoring and making lifestyle and dietary therapies to support your cervical health can have a tremendous impact. On or off the pill, begin including antioxidants like vitamins C and E and add decaffeinated green tea in your diet or as supplements. Antioxidants have been shown to have anticarcinogenic properties and to be beneficial to women with cervical changes. Cruciferous vegetables, such as broccoli, cauliflower, and kale, contain phytonutrients and support healthy estrogen metabolism. I recommend indole-3-carbinol (I3C) and diindolylmethane (DIM) supplements to my patients because they offer

Am I at Risk for Cancer?

You have an increased risk of cancer if you:

- have a family history of cancer
- are exposed to chemicals and cancer-causing substances (carcinogens)
- smoke
- drink alcohol
- have chronic inflammation
- eat a poor diet
- are obese
- have infections known to cause cancer
- have been exposed to radiation

higher doses of phytonutrients than what most of us would eat normally. (If you want to eat 2 pounds of these vegetables a day, every day, more power to you!) I3C and DIM help you convert 16OHE1 to the more beneficial 2OHE1 form.

The Brighten Metabolic Protocol

If you want to balance your blood sugar to help prevent the metabolic mayhem of insulin resistance, elevated cholesterol, high blood pressure, and hormonal imbalance, a number of lifestyle changes can help you avoid going down a road that may lead to increased risk of stroke, heart attack, or cancer. The 30-Day Brighten Program will help reduce your risk of metabolic issues, but here are the key tips of this protocol to get you started—**BEAT**:

>**B**anish sugar and refined carbs
>
>**E**at real food—with plenty of veggies
>
>**A**ctivity daily
>
>**T**imed meals

Banish Sugar and Refined Carbs

If quitting sugar makes you feel a bit like you're stuck in an intervention, then know you're not alone. Sugar is highly addictive. I've talked about this in earlier chapters, so I won't repeat myself too much here, but the key to balancing your blood sugar and sustaining your energy lies in not consuming sugar and refined carbohydrates, such as white bread, rice, and pastries, which lead to those blood sugar spikes and energy dips. I'm lookin' at you, bagels and cookies. Use maple syrup, honey, and coconut sugar sparingly. Aim for fruits like berries.

Eat Real Food—with Plenty of Veggies

Whole organic food—meat, fish, vegetables, and healthy fats—will nourish your body and stabilize your blood sugar, unlike processed foods, which often cause it to spike. I recommend you consume fat and protein at every meal for balanced blood sugar levels. Try to have 1 to 2 tablespoons of healthy fat with each meal, such as coconut oil, olive oil, avocado oil, macadamia nut oil, or whole avocado, or grass-fed butter and grass-fed ghee if dairy works for you (follow the Brighten Protocol diet to find out). For protein, aim to have 6 ounces of fish or poultry, 4 ounces of red meat, or ¼ cup of nuts.

Fill at least half your plate with fiber-rich vegetables. Avoid too many starchy vegetables, like potatoes, squash, and peas. While cruciferous vegetables in general are amazing for their anti-cancer compounds, broccoli sprouts are the bomb. They're extra potent and contain strong amounts of sulforaphane, which supports hormone detoxification and has anti-cancer properties. Broccoli sprouts have even been shown to help you eliminate environmental toxins. If you're on the pill, eat ¼ cup three or four times weekly. If you're off the pill, aim for at least twice weekly. If you have symptoms of estrogen dominance, eat ¼ cup three times per week and then daily five days before starting your period.

Aim for at least seven to nine servings of vegetables per day. If this sounds like a lot, start by adding a bit more each day. Throw a few veggies in with your morning eggs, have a large salad with protein for lunch, and include a variety of vegetables with your dinner at night. Or make one of my smoothie recipes (see page 291) for an easy way to get in a few servings at once. Try to eat a spectrum of color so that you obtain a variety of antioxidants and nutrients.

Activity Daily

Both cardio and strength training have been shown to improve glucose metabolism and insulin sensitivity, which improves hormone balance and helps you avoid metabolic mayhem. Research has also shown that regular daily exercise is effective in lowering the risk of many cancers. Moderate to vigorous exercise, about 150 minutes weekly, is associated with lower breast cancer return rates and death from the disease. In fact, a meta-analysis found that regular exercise may reduce the risk of death from breast cancer by 40 percent! Aim for at least 20 minutes of daily movement—you can alternate between cardio and strength training. Or try yoga, Pilates, Zumba, Barre3, or martial arts. If you aren't used to regular exercise, start out with walking and stretching, and work your way up to other, more intensive activities.

Timed Meals

If you experience hypoglycemia or are coming off the pill, aim for three regular meals a day—breakfast, lunch, and dinner—because you don't want to be in a situation in which you're hungry and you're losing energy and you grab a quick carb-laden snack that causes blood sugar dysregulation. If you do find yourself needing a snack even with three meals a day, have a vegetable with a protein or healthy fat, like carrot sticks and guacamole or celery and hummus. Do not reach for the pretzels or chips! If you're often on the run, especially at breakfast, make a green smoothie to take with you or a hard-boiled egg. I also recommend keeping quick, healthy snacks available, like Epic Bars, Exo bars, Paleovalley sticks, hydrolyzed collagen travel packets, or Rxbars. These ward off that hangry monster.

Intermittent fasting has been shown to be beneficial in lowering cancer risk, improving heart health, aiding in gut mobility, lowering inflammation, stabilizing blood sugar, and aiding in weight loss. Consider it your daily metabolic love sesh. I recommend women begin by aiming for a fourteen-hour fast daily. Fourteen hours! I know—it sounds like a lot, but it is easy. Promise. Aim to eat your meals within a ten-hour window each day, then close the kitchen down around 6:00 or 7:00 p.m. Let's say you eat dinner at 6:00 p.m.; the next day wait to eat breakfast until 8:00 a.m. There are many variations of intermittent fasting, but I recommend working with a doctor before experimenting beyond this basic fast. I recommend doing this as part of the 30-Day Brighten Program.

Key Takeaways: Reverse Metabolic Mayhem

- Women with PCOS often have insulin resistance, which can stimulate the ovaries to secrete testosterone.

- If you have been on the birth control pill, you can have symptoms of PCOS that are not *real* PCOS but instead post-pill PCOS or pill-induced PCOS.

- The pill has been shown to cause elevated cholesterol, blood pressure, and insulin resistance, which can lead to diabetes.

- The pill also increases the risk of blood clots, which in turn increase the risk of stroke.

- If you have the factor V Leiden or *MTHFR* gene variations, you may be at greater risk of stroke if you're on the pill.

- Women who experience auras with their migraines are at an increased risk of ischemic strokes.

- Certain birth control pills, like the kind that contain drospirenone, may increase the risk of heart attack and stroke.

- The pill increases the risk of certain cancers too, such as breast, cervical, liver, and brain.

CHAPTER 9

TAKE CHARGE OF YOUR MOOD SWINGS, ANXIETY, AND DEPRESSION

Did you know that the pill may interfere with your ability to respond to fear and stress in an appropriate way? Did you know that HPA axis dysregulation (which women who take the pill have, as we talked about in chapter 7) can put you at the mercy of stress, giving you altered fear responses? Did you know that it's possible for the pill to influence how you interact with your children and even your mate? Research has also shown that women who use the pill have decreased sensitivity to oxytocin, which can impact their ability to bond with their baby or their partner. It affects relationship satisfaction as well—and not in a good way. This all makes sense, because research has shown that the natural cyclical changes of hormones influence mood, behavior, and social bonding. It's an area that's just starting to be explored. Many researchers are now starting to examine how the birth control pill may be influencing not only your mood but also how you socialize and behave as a person altogether.

Like many women, you may have been told that a connection between the birth control pill and your mood is just all in your head. It's commonplace not only in medicine but also in society as a whole to dismiss the mood symptoms associated with the pill. Never mind that the symptoms are listed as side effects right on the packaging. Women have been reporting altered mood since the pill was introduced. But depression isn't the only mental health challenge the pill poses. There's also a connection between the pill and a somewhat less visible but far more pervasive disorder: anxiety.

If you suffer from depression or anxiety—or both—you may feel desperate and damaged, but what I want you to know is that you're not broken. And if your doctor has told you that it's "all in your head," then it's time to get a second opinion. For many women, starting the pill means a decline in their mood, but there are many natural ways to support your body in healing from these devastating, at times paralyzing disorders that can arise when you begin taking the pill. In this chapter, I'm going to share with you some effective and proven approaches to help restore your mood—and kick depression and anxiety for good!

Did Your Mood Tank After You Started the Pill?

Three months after stopping the pill, Samantha found her way to my office. She was struggling with her lack of motivation and, in her words, she had "fallen out of love with life." She found herself crying easily and feeling hopeless every day. Her symptoms had begun four years prior, when she had started the birth control pill. And after many years of her doctor telling her that there was no way the pill had anything to do with her depression, she decided it was time to call it quits and see for herself. Samantha had thought, like many women, that as soon as she kicked that pill, her mood would come right back and she'd be feeling like her old self again. However, she wasn't aware of post–birth control syndrome and the long-term side effects and consequences the birth control pill can create. She couldn't understand why her depression persisted, and it honestly began to make her doubt herself. *Maybe her doctor was right. Maybe it didn't have anything to do with the pill.* I assured Samantha that she wasn't the first woman to report mood changes with starting the pill, and she certainly wasn't the first woman to come to my office after struggling despite having quit the pill. I explained to her that there are some very real ways the pill can impact your mood, and there was a root cause as to why her mood had not lifted and instead she suffered from constant feelings of depression.

In a thirteen-year study that followed more than one million women aged fifteen to thirty-four, researchers found the women were more likely to be

In This Chapter

- How the pill can make you feel depressed

- The two primary risk factors for developing depression while on the pill

- How the pill interferes with your serotonin production

- The connection between anxiety, your hormones, and the pill

- How the pill causes nutrient deficiencies—and what you can do about it

diagnosed with depression for the first time if they used hormonal contraception. These women were 23 percent more likely to be prescribed an antidepressant if they were taking a combination pill that contained both estrogen and progestin. Even more startling, teens taking a combination pill were *80 percent more likely to develop depression.* Unfortunately, the progestin-only pill wasn't much better for teens, increasing their risk twice as much as those not taking a birth control pill. This study was groundbreaking for women's health because it was the first time a study of this scope revealed a conclusive relationship between hormonal birth control and depression. While prior studies had never been able to demonstrate a definitive link, this one finally had exposed the real risk that taking birth control can lead to a diagnosis of depression. It provided much-needed insight for women—and doctors!—who have been grappling with or unaware of this connection for years.

Dr. Kelly Brogan, a board-certified psychiatrist and author of *A Mind of Your Own*, believes the pill represents a significant obstacle to mental health and appropriate hormone balance. Brogan explains, "These days when I meet a patient who complains about flat mood, low libido, weight gain, irritability, depression, and anxiety, one of the first questions that I ask is 'Are you on the pill?' And it appears that there is a subset of patients for whom synthetic hormones are a really bad fit and can exacerbate either preexisting psychiatric symptoms or manifest new psychiatric symptoms." Brogan believes in investigating your symptoms, acknowledging, "I have come to the perspective that you can never truly own your primal femininity, never answer the question of what your symptoms are asking—from irregular periods to PMS—without deep contact with your hormonal self."

Increased Suicide Risk

Another recent study has discovered that young women who use hormonal contraceptives have *three times the risk of suicide* compared to those who have never used this type of birth control. According to this study, the threat of suicide was highest during the first two months after beginning the pill (or the ring, IUD, or patch). While this risk did plateau after a year, it still remained higher compared to that of women who never used hormonal contraceptives. The patch had the highest danger of suicide attempts, with the other hormonal contraceptives following closely behind. No hormonal contraceptives are without risks. While research studies can certainly have many variables that make it challenging to prove direct causation, these findings should make us reconsider how freely we prescribe hormonal contraceptives—especially for reasons other than birth control.

I believe in a woman's right to prevent pregnancy however she chooses, but knowing the risks associated with the pill, as well as how to protect yourself from these risks, is important to maintaining your health. The new feminism includes being so well informed about your body that you can be 100 percent confident that you're making the absolute best decision for *you*. If you're going to choose the pill or another hormonal contraceptive as your primary form of birth control, you need to be completely informed of the mental health risks, aware of the potential signs and symptoms, and armed with the tools that can help you counteract the detrimental effects. I've been advocating for years for more thorough screening and personalized counseling by doctors before they recommend hormonal contraceptives to their patients, because understanding each woman's individualized needs can help us make the recommendations that are going to serve her and her health best.

Are You at Risk?

Why do some women experience depression when taking the pill while others don't? Well, the short answer is that every woman's body is different—and this is why root-cause medicine that looks at each person individually is so

The Pill and Depression Risks

- Women taking combination pills (estrogen and progestin) were 23 percent more likely to be prescribed antidepressants.
- Teens were 80 percent more likely to develop depression when taking the combination pill.
- Women taking the progestin-only pill were 34 percent more likely to be prescribed an antidepressant.
- Teens taking the progestin-only pill saw a two-fold increase in their risk of depression.
- Young women who use hormonal contraceptives have three times the risk of suicide.
- Teens have double the risk of suicide after one year on the pill and are at a 30 percent higher risk after seven years on hormonal birth control.
- Suicide risk peaks around two months after beginning hormonal birth control.

important. We do know of two risk factors: a personal or family history of depression and a personal or family history of immune dysregulatory diseases or inflammatory diseases like autoimmune disease. That means if you've experienced depression in the past but have since resolved it, it may return while taking the pill. Or if your mom, brother, or grandparent had depression, you may be more susceptible.

Research has shown that inflammation and immune system dysregulation are involved in depression. As you learned in chapters 6 and 8, the pill is inflammatory and leads to a compromised immune system. Because of this, a personal or family history of inflammatory or autoimmune disease may place you at greater risk of experiencing depression while on the pill. This is an area in need of more attention and research to deepen our understanding.

For my patient Samantha, while she herself had no history of depression before taking the pill, she did recall that her mother had struggled with postpartum depression. We also found clues in her lab work as to why she was feeling depressed: she had elevated inflammatory markers and low levels of B6. As we'll explore in more detail, increased inflammation can raise the risk of depression.

How Can the Pill Contribute to Depression?

The pill depletes nutrients crucial for brain health, lowers testosterone, disrupts your thyroid and adrenals, causes leaky gut, and messes with your microbiome—any one of these alone can cause you to feel depressed. Research is now showing women on the pill experience a decrease in the neuroprotective molecules that protect brain cells and an increase in the neurotoxic chemicals that destroy brain cells compared to women who do not use hormonal contraception. That means being on the pill is bad for your brain, and it appears your mood depends on how one little amino acid—tryptophan—gets processed.

Brain-Supporting Nutrients

The following will help replenish nutrients and support your brain:

- acetyl-L-carnitine
- *Bacopa monnieri*
- berries
- beta hydroxybutyrate
- coffee
- epigallocatechin gallate (EGCG)
- fish oil
- Ginkgo biloba
- huperzine A
- magnesium
- phosphatidylcholine
- phosphatidylserine
- turmeric

What's Turkey Got to Do with Your Brain?

Have you ever heard that turkey makes you sleepy? Well, that's because turkey contains tryptophan, an amino acid you use to make serotonin (a chemical found in the brain and gut) and melatonin, the sleepy neurotransmitter. While that turkey business is a myth (it's really all the carbs, sugar, and alcohol you consume on Thanksgiving that make you sleepy), tryptophan is an essential amino acid in brain, hormone, and mental health. When tryptophan metabolism is functioning optimally, your body produces serotonin and melatonin as well as kynurenic acid, which protects your brain. Studies have shown that women on the pill don't metabolize tryptophan normally and some have recommended daily tryptophan supplements to rectify this.

One study found that women on the pill had reduced kynurenic acid and elevated high-sensitivity C-reactive protein (hs-CRP), a sign of inflammation and immune system activation. (Remember, the high dose of estrogen in the pill is inflammatory.) When inflammation or cortisol climb, the tryptophan pathway shifts toward quinolinic acid production, which is inflammatory and harmful to the brain. Inflammation and immune dysregulation have been shown to play a role in the development of depression. For my patient Samantha, it definitely appeared as if inflammation was the root cause of her mood symptoms. Furthermore, the pill depletes B6, which is necessary for converting tryptophan to serotonin and kynurenic acid, and Samantha's labs had indicated low levels of B6.

We know that hormones affect the immune system—there are estrogen receptors on key immune cells, like lymphocytes, macrophages, and dendritic cells. We also know that many genes in the innate immune system respond to estrogen. Research has shown that women typically mount a stronger immune response to infection and are more likely to suffer from inflammatory and autoimmune disease than men. This is a major reason researchers believe women have a higher risk of depression compared to men.

Okay, so after knowing all of this, it makes perfect sense that an estrogen-containing pill can influence an inflammatory response and shift the body to make harmful brain chemicals. But there appears to be something special about the pill itself, since estrogen also has been shown to benefit the brain and protect against neurological diseases. Because of this, many experts hypothesize that the multifactorial effect of the pill, including the increase in inflammation and nutrient depletions, may be a primary reason women report depression while taking the pill. Larger studies are needed to understand if reduced kynurenic acid and increased inflammation are the primary mechanism of how the pill may cause depression.

Never Dismiss Your Risk—or Symptoms!

If you are on the pill and struggling with depression, I recommend taking a good look at what role that pill may be playing in your overall mood and considering breaking up with it. I'd love to tell you it will all just magically get better once you do, but unfortunately that is not what I've found to be true clinically. It may take some time, changes, and commitment to achieve hormonal health. If all you do is quit the pill, you shouldn't expect your mood to turn around. If you quit the pill, know that only a small percentage of women feel better when it comes to their mood symptoms. And the majority of women need to employ diet and lifestyle therapies along with targeted supplementation to really reclaim their mood. So if you're currently having mood symptoms, taking the steps in this book right now can help to improve that, and it will make it a whole lot easier to transition off the pill's hormones once you're ready to do so. And I want you to know that Samantha was able to take back her mood with the information I'm sharing in this chapter.

We began with the 30-Day Brighten Program diet, which is inherently anti-inflammatory and supplies the brain with the fats essential to brain health; used a turmeric supplement to reduce inflammation; and supported Samantha's adrenals so her body's natural anti-inflammatory mechanisms could be leveraged. We also began to restore her microbiome using the Brighten Gut Repair Protocol and probiotics. There's been an explosion in microbiome research showing that what grows in your gut absolutely influences how you feel throughout your day. Within one cycle, Samantha's mood lifted, and she felt a lot better the week before her period. After three cycles, she was happy to report that she had found her groove once again and felt highly motivated at her job and in life altogether. She followed up ten months later, and it was almost as if she had never experienced the depression to begin with. She felt so full of joy and energy for life that she'd put the whole ordeal behind her. She also worked with a counselor, because if someone has depression or anxiety, I make sure she has a mental health specialist working with her. I think it's really important that women have access to experts in each arena to be able to meet all their needs.

The Pill and Nutrient Depletions

Research has shown that taking the birth control pill depletes your body of important nutrients. Here's a list of some of the common nutrient depletions you may experience:

- folate (folic acid)
- magnesium
- selenium
- vitamin B2 (riboflavin)
- vitamin B6 (pyridoxine)
- vitamin B12
- vitamin C
- vitamin E
- zinc

In addition, the pill lowers antioxidants, such as coenzyme Q10 (CoQ10). Because of all of these depletions, it's important to consume a nutrient-dense diet and begin taking supplements, including CoQ10 and those I list in the Brighten Supplement Protocol (page 246).

If you have a history of depression or a first-degree family member with depression, then I urge you to consider an alternative contraceptive. While the debate about whether the pill causes depression continues on, it has been well recognized in medicine that a personal or family history does place you at risk for mood changes when you start the pill.

I cannot stress enough that if you are on the pill and have mood symptoms, don't let anyone, not even your doctor, dismiss your symptoms or brush them off as being "all in your head." Depression is a symptom of imbalance, a way your body communicates to you, and it can be downright debilitating. You deserve to have your symptoms investigated.

If for whatever reason you need to or choose to stay on the pill, make sure you are taking steps to replenish nutrients depleted by the pill, to reduce inflammation, and to support your liver in natural detoxification. Begin taking a prenatal or multivitamin and follow the Brighten Mood Mastery Protocol as well as the liver and inflammation supportive protocols in chapters 5 and 6. The 30-Day Brighten Program in chapter 12 will help enormously with revitalizing your mood.

Is the Pill Giving You Anxiety?

Maybe you don't suffer from depression but you pop a Xanax a day to keep the anxiety away. While the thought of an unplanned pregnancy may stress you out, taking the birth control pill may ultimately cause far more anxiety because of what it does to your body. If you're experiencing debilitating anxiety levels, there's a good chance you have a hormonal imbalance—especially if you've noticed a correlation between your symptoms and when you began taking the birth control pill.

Anxiety is no joke. It can be incredibly debilitating when you have a sense of fear and panic that seemingly comes out of nowhere, and it can seriously impact your life. In my clinic, if a woman says that anxiety is her primary symptom, the first thing I want to do is get her symptom relief quick so she can take all that energy being eaten up by her anxiety and refocus it on actually healing herself and working on the root cause.

There is a real connection between anxiety and your hormones—and no amount of medication or meditation is going to help alleviate your symptoms if you don't get to the root cause of why you're having anxiety. The conventional belief is that anxiety is caused by a neurotransmitter imbalance, and many doctors will prescribe a pharmaceutical to try to shut down all that bad behavior. But while medication may suppress the uncomfortable symptoms, it does nothing to rebalance the neurotransmitters, fix the gut, or address any of the underlying causes of anxiety.

The truth is that a lot of anxiety is caused by stress, which, as you now know, generates a response from the adrenals to produce more cortisol. In today's world, most of us experience some form of stress. Every. Single. Day. Whether it's an important client meeting or a looming work deadline, a fight with your significant other or an issue at your child's school, these repeated stressful situations launch a response in your body that can spark anxiety, especially if you're on the pill. Besides the more obvious causes of stress, subtler stressors constantly bombard the body too, such as exposure to light at night, eating on the run, or chronic infections.

First, let's take a look at how stress affects your cycle. Around day 14, once your egg is released during ovulation, your body begins to secrete progesterone for about two weeks. Progesterone induces a sense of calm and a deep sense of love and connection by stimulating gamma-aminobutyric acid (GABA) receptors, which quiet the excitatory (i.e., stress!) neurotransmitters. Unfortunately, when you're under stress, your body prioritizes survival over fertility, which means it begins making more cortisol at the expense of progesterone. When you suffer from chronic stress—as many of us do—you experience HPA axis dysregulation, which sets off the adrenal fatigue symptoms you learned about in chapter 7. Your body ramps up the epinephrine and norepinephrine, which tell your brain to freak out. Since your body also drastically reduces your progesterone, which would normally instruct your brain to remain calm, you have double the impact of stress and anxiety. You know what else disrupts the HPA axis? Inflammation. And the birth control pill is definitely a major instigator of inflammation. Remember, the pill also shuts down ovulation, which means you don't make progesterone while on it, and progestin does not have the same mood benefits.

Anxiety can and often does originate from hormonal imbalances like those caused by HPA axis dysfunction. Women with Hashimoto's thyroiditis are twice as likely to develop anxiety (see chapter 7 if this is you). It also stems from compromised gut health and nutrient deficiencies, as I touched upon in chapter 6. Your gut actually creates a significant amount of serotonin and is therefore incredibly important to keep healthy if you want to avoid anxiety. Nutrient deficiencies place you at a higher risk for anxiety, and nutrient-depleting medications like the pill exacerbate any deficiencies you may already have. The following nutrients all affect your mood and can lead to anxiety when you have a deficiency:

- copper
- EPA and DHA
- folate
- inositol
- magnesium
- omega-3 fatty acids
- vitamins B6, B12, and D
- zinc

So if you're on the pill and have noticed that you're more anxious than usual, consider whether there is a correlation between when you began the pill and your changes in mood. And whether or not quitting the pill is an option, you can begin taking action now to reduce these paralyzing symptoms and restabilize your mood with my Mood Mastery Protocol.

The Brighten Mood Mastery Protocol

You now know the gut and immune system play a pivotal role in your brain's health, and the birth control pill depletes a whole host of nutrients vital to that health. The 30-Day Brighten Program in chapter 12 will help you rid your body of toxins and inflammation, rebuild your gut, and replenish nutrients. In the meantime, here are a few essentials to alleviate depression and anxiety, by reducing inflammation and correcting nutrient depletions with nutrient-dense foods and supplements. The key steps to the Brighten Mood Mastery Protocol are **EMBODY**:

> **E**at for your mood
>
> **M**ove your body
>
> **B**anish stress
>
> **O**wn your sleep
>
> **D**aily smart supplements
>
> **Y**our lady tribe

Eat for Your Mood

If inflammation is contributing to your depression or anxiety, it may be due to digestive disorders, autoimmunity, food sensitivities, stress, chronic infections, the birth control pill, or more. A qualified physician can help you determine the root cause, but for now you can begin healing with an anti-inflammatory diet. Sugar and food sensitivities can increase anxiety and leave you feeling less than fabulous. Eliminate all grains and gluten, dairy, soy, sugar, caffeine, alcohol, and inflammatory fats (found in processed or refined foods, fast foods, margarine, canola oil, corn oil, cottonseed oil, and peanut oil). You can then slowly reintroduce some of these foods one at a time to understand which may

Why Fat Can Boost Your Mood

Healthy fats stabilize your blood sugar, reduce inflammation, regulate hormones, and fuel your brain. Your brain needs fat—in fact, 60 percent of your brain is fat. If your diet is too low in fat, it affects your memory and mood. Fat is rich in vitamins A and E, and omega-3 fatty acids, which can help reduce inflammation when you have depression and boost your brain power. Because fats fill you up, they regulate your blood sugar so you can avoid the blood sugar spikes that can contribute to anxiety. Blood sugar dysregulation can spike cortisol and exacerbate symptoms of worry and fear. If you're trying to counteract symptoms of depression or anxiety, add some healthy fats to your diet.

be contributing to your mood issues. See chapter 6 for more specifics about gut healing and chapter 12 for detailed information on food reintroductions.

Begin eating whole foods to replenish your nutrients and reduce inflammation. Opt for more vegetables and healthier fats. Omega-3 fatty acids can help with your inflammation while also feeding your brain, and you can find them in cold-water fish, like mackerel, salmon, and anchovies, or take fish oil with a mix of EPA and DHA. Turmeric is also a strong anti-inflammatory; you can steep the fresh root for a tea or take it as a supplement. Incorporating herbs into your diet can also provide relief, and many have long been used to treat depression and anxiety (see more in the following "Daily Smart Supplements" section).

Move Your Body

Moderate exercise can lower the inflammation that messes with your mood and increase brain function, not to mention give you an endorphin high. It also protects you from developing neurological conditions like depression and dementia. Moving your body every day does wonders for your mood and helps regulate your hormones. Find exercises you enjoy, such as Zumba class, yoga, Pilates, walking, or weight training, and commit to regular activity. You may want to step up your intensity a bit because when you're anxious, your body wants you to move so that you can relieve tension and counteract the effects of cortisol. Big muscle movements like jump squats, lunges, and kicks can help you move through anxiety and reduce stress. Speaking of stress . . .

Banish Stress

Since you already know that stress causes you to crank out cortisol, imbalances your hormones, and triggers inflammation, which leads to anxiety and depression, you definitely want to become more aware of what causes your stress and take steps to eliminate it. See the stress-reducing practices listed on page 260 of the program for ideas.

You also may want to nix the caffeine if you're constantly feeling stressed (I recommend cutting out caffeine completely during the 30-day program). That morning java jolt could be doing more harm than good. Okay, I *love* coffee, but I gotta tell you that many of my patients notice a decrease in their anxiety when they stop drinking their morning cup. If you relish morning coffee, then I suggest gradually tapering off with half regular, half decaf. You can also replace your daily coffee with decaffeinated green tea, which contains L-theanine, a calming amino acid. If you're like me and love the taste of coffee, try one of the caffeine-free coffee replacements on page 122.

Own Your Sleep

I've said it before and I'll say it again: sleep is absolutely critical to your hormone health. If you're not sleeping at least seven hours per night, you can kiss those awesome moods goodbye. Your brain and body recover from the day during sleep, so if you're not getting adequate sleep, you can expect your body to struggle. It's important to reinforce your circadian rhythm to boost your mood. See "Reset Your Circadian Rhythm," page 233, for some helpful tips.

Daily Smart Supplements

Even when my patients are consuming an ideal diet, I often include supplements for those who either are or were on the pill in order to optimize their nutrient intake. I also prescribe supplements for those who are in a healing phase and are taking steps to diminish and eliminate unwanted symptoms. Supplements can help reduce inflammation and support optimal brain health, and they are a great way to replenish the nutrients lost by taking the pill. Many researchers have long documented the nutrient-depleting effects of the pill and recommend a prenatal or multivitamin as a first line of therapy. Hormone-balancing herbs and nutrients can also help maintain healthy progesterone levels.

In addition to a quality multivitamin, I recommend taking the following:

- **Turmeric** at a dose of 1,000 milligrams once or twice daily (also see the recipe for Upgraded Golden Milk," page 304)
- **Omega-3** at a dose of 2,000 to 4,000 milligrams daily
- **B-vitamin complex**
- **Passionflower** is a lovely little plant that can help alleviate anxiety. Try taking 2 or 3 dropperfuls of tincture when you're feeling anxious.
- **L-theanine** is an amino acid found in green tea and known to produce a sense of calm in your body. Take 200 milligrams twice a day.
- **Taurine** is a precursor of GABA, which many people with anxiety are low in. Take 500 milligrams daily.
- **Glycine** is an amino acid in collagen that helps you feel calm. Eat 2 to 4 tablespoons of collagen daily.
- **Magnesium** at a dose of 300 milligrams nightly
- **Vitamin D** may be something you're low in, but get tested first. Take 2,000 to 5,000 IU daily depending on your status.

Botanicals such as passionflower and skullcap support healthy serotonin production by reducing quinolinic acid production. Adaptogenic herbs like licorice root, Rhodiola, holy basil, and ashwagandha can improve your cortisol production and lower inflammation. While the herb St. John's Wort has been shown to be beneficial for mood, it also affects your liver in a way that can make the pill fail. For this reason, it is recommended that women on the pill do not take St. John's Wort.

Your Lady Tribe

Women need community, and when we're with our tribe, we actually have better progesterone levels. Not only that, social isolation—especially as we leave college and enter the workplace or become mothers—can really weigh on women, so spending time with friends who support us can help our mood and our hormones. If you're struggling with depression or anxiety, it's important that you have adequate support, and while you'll employ all the lifestyle and dietary recommendations in this book, you also should work with a mental health expert to make sure you're taking a holistic approach to your needs.

Key Takeaways: Take Charge of Your Mood Swings, Anxiety, and Depression

- Studies have confirmed that women taking hormonal contraception are more likely to be diagnosed with depression.

- Teens taking a combination pill were 80 percent more likely to develop depression.

- Young women taking hormonal contraceptives have three times the risk of suicide—with the greatest threat occurring within the first two months.

- Two well-known risk factors for developing depression while on the pill are a personal or family history of depression and a personal or family history of immune dysregulatory diseases and inflammatory diseases like autoimmune disease.

- The pill depletes nutrients crucial for brain health, lowers testosterone, disrupts the thyroid and the adrenals, causes leaky gut, and messes with the microbiome, all of which can cause depression.

- The pill interferes with tryptophan metabolism, so instead of producing serotonin, melatonin, and kynurenic acid, which protect the brain, it elevates hs-CRP and shifts toward quinolinic acid production, which is inflammatory and harmful to the brain.

- There is a real connection between anxiety and hormones, and the pill causes the HPA axis dysregulation, compromised gut health, progesterone blockage, and nutrient deficiencies that can trigger anxiety.

- The birth control pill depletes vitamins B, C, and E, magnesium, selenium, zinc, and coenzyme Q10, all of which can impact your mood.

- To boost your mood and combat depression and anxiety, take the following key steps: eat for your mood, start taking smart supplements, get moving, banish stress from your life, obtain adequate sleep, and embrace your tribe.

CHAPTER 10

BOOST YOUR LIBIDO AND FERTILITY

Robbing you of your libido may be the most effective way the pill actually works, which is kind of sad and also frustrating, because you start taking it so you can be sexually active without having to worry about having a baby. But once you get on it, you find yourself not interested in sex in the least. You may have heard the pill can ruin your libido, but this often gets brushed off as some kind of urban legend. It's not. It's a real thing that happens to women, and it's just lame. Plus, so many of us turn that finger around and point it right at ourselves, thinking, *There must be something wrong with me. Or maybe it's that I'm just not into this guy anymore.* Or maybe you start making excuses like, *Hey, this is just how it goes in relationships. It's just not exciting like it used to be.* Or *I'm just getting older.* While there might be some truth to that conversation you're having with yourself, odds are, if you're on hormonal birth control, that's what's really robbing you of your libido. And if you're a gal who's thinking about having kids, you're probably sitting back right now and wondering, *Wait, don't I at least have to want to have sex to have a baby?* Yeah, that is how it typically works, just in case you needed some reassuring!

On top of being a you-know-what blocker, the pill can seriously mess with your fertility. Most women are told that the pill has very little impact on their fertility and that when they get off it they'll be able to become pregnant easily:

"It's nothing to worry about." If you're someone who's not interested in making a baby any time soon—or ever—then you may be wondering why in the heck we're talking about sex and fertility in the same chapter. Here's a little secret I'm going to let you in on: a fertile female body is one that has a robust libido and killer orgasms. So whether or not you want a baby, I got you here. We're going to get your libido back, we're going to get your orgasms back, and when it's time to have a baby—if that's in your cards—we're going to make sure your fertility is taken care of as well.

Doctor's Orders: Orgasm Once Weekly

Women can have four different types of orgasm throughout the month because of the hormonal changes we experience. Dude, this is a serious reason to consider ending that pill pack. Believe it or not, I have to make the case for why women should have orgasms not only to other women but also to my professional colleagues. Because—no joke—how many doctors are like, "Orgasms are nice, but they're not really necessary for women"? That's a straight-out lie. Besides feeling amazing, orgasms have incredible health benefits. They can reduce stress and anxiety, increase circulation, and improve autoimmune disease, menstrual cycles, and fertility. They also can give you glowing skin, relieve migraines, and result in better sleep. Orgasms can even help you live longer! So if you're not currently enjoying regular orgasms, start making it a priority. Doctor's orders!

When you have an orgasm, you release oxytocin, often called the "love hormone" because it increases your feelings of affection or bonding with your partner. It also promotes social bonding in general. Oxytocin can counter the negative effects of cortisol, so it can help relieve stress. It can even sharpen your intuition, and it may be why some women are so successful. See? Orgasms are even good for your career! Research has shown that oxytocin also can reduce anxiety and social anxiety.

Most of us spend way too much time sitting—whether that's at a desk at work or in front of a television at night. So much sitting is not good for the body and can even lead to decreased circulation in the pelvis. Orgasms can increase circulation to the pelvic floor organs, helping nutrients and hormones get where they need to go. Hula-hooping, hula dancing, and belly dancing can help too, but let's be honest: orgasms are way more fun.

In This Chapter

- The extraordinary health benefits of orgasms

- How the pill can hijack your libido—and what to do about it

- The impact the pill can have on your fertility

- Why you should optimize your health before trying to conceive

- The top seven foods for maximizing your libido

A low libido is common in autoimmune disease, but orgasms can be beneficial for immune health. For simplicity's sake, let's focus on two aspects of the immune system, Th1 and Th2. Th1 is the part of your immune system that fights off viruses and bacteria, and anything that is "not you." This could potentially even include sperm or a baby if you were to conceive. Th2, however, is much more tolerant, which is why pregnant women experience a shift in their immune systems from Th1 to Th2. Th2 is the aspect of the immune system that developed to protect against parasites, although in modern times it presents more as allergies, asthma, and eczema. The majority of autoimmune diseases are driven by the Th1 system, which increases autoimmune symptoms. Th2, on the other hand, can decrease these symptoms, so if you're suffering from an autoimmune disease, you want to shift Th1 and Th2 into a balanced state, which orgasms can help you do. Sex can actually decrease autoimmune symptoms and boost fertility, because the immune system shifts to a Th2 state, which is more favorable for conception. Women who are not having sex do not experience this immune shift. What if you're Th2 dominant? Orgasms FTW again, because they harmonize your overall immune system.

Research has shown that women who have sex on a weekly basis also have more predictable menstrual cycles. Women who don't have sex regularly may have more sporadic cycles, which tend to be shorter, a possible indication of low progesterone and estrogen dominance, but regular orgasms can have a hormone-balancing effect. They also can help relieve menstrual cramps, because releasing oxytocin and other endorphins during orgasm may reduce pain.

Orgasms improve your overall fertility not only because they give you more regular cycles but also because women who have sex weekly have the highest incidence of fertile basal body temperature (BBT) rhythms. Orgasms also shift you into that Th2 immune system state at the right time, which elevates your chances of conception.

Because orgasms increase your circulation, they nourish your skin, giving you a gorgeous glowing complexion. Next time you have sex, take a look in the mirror afterward—that glow will already be visible. Orgasms release anti-inflammatory chemicals that protect your skin from environmental toxins and pro-aging hormones. Who knew they could be such an important part of your beauty regimen?

Did you also know they can offer relief from migraines? The pain-reducing and relaxing effects of the hormones you release during orgasm can help eliminate those stubborn headaches. So next time your partner wants to have sex and you respond, "Not tonight. I have a headache," you may want to think twice and get busy in the bedroom.

The Top 10 Health Benefits of Orgasms

1. You will live longer. (Really, I could have just stopped this list at number 1.)
2. You will look and feel younger from the antiaging hormones released during orgasm.
3. They decrease your autoimmune symptoms.
4. They improve your mood and reduce anxiety.
5. They increase circulation to the pelvis.
6. They decrease stress and promote relaxation.
7. They give you easy periods, more regular cycles, and relief from menstrual cramps.
8. They provide relief from migraines.
9. They boost your fertility when you have them regularly.
10. They help you sleep better.

Trouble with anxiety? Have sex. Trouble with insomnia? Have sex. When you have an orgasm and release oxytocin, it relaxes you, calms your mind, and enables you to get a good night's sleep. Besides oxytocin, your body releases vasopressin during orgasm, a hormone that often accompanies the release of melatonin. As you know, melatonin is the "sleep hormone." Ever wonder why men often fall asleep right after sex? Now you know.

Finally, orgasms can even help you live longer. DHEA is an antiaging hormone that begins to decline in our mid to late twenties. But orgasms help increase your DHEA, which also improves brain health, skin, and immune function—all important for combatting the effects of aging. DHEA can also reduce autoimmune antibodies and act like a natural antidepressant. Best of all, research has shown that people who have regular orgasms have a 50 percent reduction in mortality risk. Having frequent orgasms really can save your life.

Bye-Bye, Libido!

Erin felt major chemistry with Jake, a guy who lived in her dorm, but she couldn't act on her feelings because he was still dating his girlfriend from home. Eventually, she started dating Andrew, a lacrosse player she met at a party one night. When they became sexually active, she decided to start taking the birth control pill. After a few years, their relationship ended, and Erin once again found herself spending time with Jake, who was now single. Finally, they were both single at the same time. The problem? Erin realized she was no longer attracted to him. In fact, she didn't want anything to do with him physically or romantically. It was a bit strange, but she chalked up her prior attraction to simply being young and new to college and didn't think much of it. And so Erin and Jake simply remained friends.

Erin was single long enough that she decided to go off the pill, because it seemed pointless to keep taking it. After being off the pill for several months, she noticed that she was super attracted to Jake again. Why was she suddenly feeling this way after all these years? One night when they were watching a movie, Jake finally kissed her and they took the leap from just being friends to having a romantic relationship. Erin believed that she had found her soul mate and was telling all her friends and family that she was going to marry him.

Because things were so serious, Erin decided to start the birth control pill again . . . and their relationship began to deteriorate. She lost her libido, didn't want to have sex with Jake, and wasn't even remotely attracted to him. She completely stopped caring about sex because it no longer felt good when she had it—she was now suffering from vaginal dryness. Also, her mood had begun to degrade. Jake admitted that he didn't recognize her anymore, because she was always depressed, crying all the time, and seemed to have lost all motivation. Before this, Erin had always been a type A, "Go get it" kind of girl, but now she felt completely unmotivated. Her doctors told her she was suffering from depression and prescribed an SSRI. She tried it and it didn't help; she didn't like how she felt because she felt nothing.

Erin finally came to see me in her mid-twenties after reading one of my blog posts, because she realized that she might be having symptoms from the birth control pill—and that maybe her depression and loss of libido were linked to a hormone imbalance. At this point, Erin felt her relationship with Jake was no longer working. She thought she had been young and stupid, and had no idea why she had ever thought she would marry him. She was seriously considering getting out of the relationship.

Before she called it quits with Jake, I helped Erin quit the birth control pill and begin the 30-Day Brighten Program. Six months later Erin was back in love with Jake and once again believed he was her soul mate. And Jake finally recognized the woman he'd fallen in love with all those years ago.

Now Erin and Jake are planning their wedding, and she's also in grad school. She's a totally different person—or, actually, she's the person she used to be before the pill hijacked her personality. Erin's story illustrates just how much the pill can mess with your libido, your mood, and your life. The pill had a major influence on Erin's attraction to the love of her life as well as her overall mood and motivation. Once we got Erin off the pill and started working on all of these things, she fell back in love with Jake, her family members recognized her again, and she was back to her ambitious self, having found her groove again. If, like Erin, your libido has gone missing in action, follow my protocol at the end of the chapter.

While many people view the female orgasm as some giant mystery, the truth is that it's a sign of health and something you—and your doctors!—should take seriously. Don't allow anyone to dismiss your concerns about losing your libido. Many women experience a diminished libido or find it difficult to achieve orgasm. I've had women of all ages admit this to me in my practice. As always, it's important to investigate the root cause, and if you've lost your sex drive, you may have low testosterone, inflammation in your body, or an HPA axis dysregulation that requires you to support your adrenals.

How the Pill Hijacks Your Libido

Unfortunately, the pill has been crashing women's libidos from the very beginning. It seems pretty counterintuitive to me that someone somewhere designed a method of birth control for women that essentially ruins any desire for sex. (I'm sure if men had to take a birth control pill, this would never have been allowed! And, of course, it wasn't, as you know from "What Happened to Male Birth Control?" in chapter 3.) Multiple scientific studies have revealed the ways the birth control pill has been detrimental to women's sexual health, including a loss of interest in sex, reduced arousal, difficulty achieving orgasms, pain with sex, painful orgasms, less frequent sex, less female-initiated sex, and overall lack of sexual enjoyment. One study of more than one thousand female medical students discovered that women who were using the pill were 32 percent more likely to experience female sexual dysfunction than those who used nonhormonal methods of birth control (or no birth control at all). These women were also 8.7 percent more likely to experience an orgasm disorder.

How, exactly, does the pill sabotage your libido? It has to do with your hormones. More specifically, testosterone and a protein called sex hormone–binding globulin (SHBG). While the pill suppresses your brain from releasing follicle-stimulating hormone (FSH) and luteinizing hormone (LH) so that you won't ovulate, it also interferes with your testosterone production. Plus, it increases SHBG to safeguard your body against all the excess estrogen the pill delivers. When you have high levels of estrogen, your body produces more SHBG to bind estrogen and protect your cells. But this protein also binds testosterone, which is already low due to the pill shutting down production in the ovaries. The result? Bye-bye, libido!

Unfortunately, your SHBG levels may remain elevated even after you discontinue the pill. A study in the *Journal of Sexual Medicine* revealed that women who were on the pill for at least six months had higher levels of SHBG than women who had never taken it, and these levels remained elevated even several months later. Women who opted to remain on the pill had about four times the normal amount of SHBG! While these levels may eventually decrease, *they will*

Lab Tests for a Low Libido

If you find that your sex drive is low, even after you've stopped taking the birth control pill, consider having the following tests:

- DHEA-S
- DUTCH Complete
- estrogen
- progesterone
- sex hormone–binding globulin (SHBG)
- total and free testosterone

never return to pre-pill levels. These findings have caused researchers to speculate if long-term exposure to the synthetic estrogen in the pill can alter a woman's genes to continue to make higher levels of SHBG for life, which of course could have a long-term effect on her libido.

Supporting Your Body Energetically

If you're having issues with your libido or trying to get pregnant, I recommend moving into your womb space. This is also known as the sacral chakra, and it is the second of the seven chakras in your body, located in your lower abdomen. The sacral chakra is the place where your creativity flows. However, this is also the place where women often store stress and trauma. You may be thinking, *But I store stress in my shoulders and my neck*, but the fact is, you can store stress and trauma in *any* tissue in your body, and the sacral chakra is a common place where we women tend to put things. Because this is also your creative center, it's super powerful. Think about it: you make babies here; you grow an entire human life from this space!

One of the reasons I became interested in women's medicine was I realized that too many women were disconnected from this space. While I was studying gynecology, I observed that women would kind of abandon this sacred territory because gyn exams were done *to* women and not *with* women. I could see that women would just scoot to the edge of the table and then disconnect from their bodies because it's not a pleasant experience. And I totally get that—I've

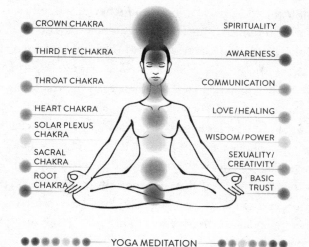

CROWN CHAKRA — SPIRITUALITY

THIRD EYE CHAKRA — AWARENESS

THROAT CHAKRA — COMMUNICATION

HEART CHAKRA — LOVE / HEALING

SOLAR PLEXUS CHAKRA — WISDOM / POWER

SACRAL CHAKRA — SEXUALITY / CREATIVITY

ROOT CHAKRA — BASIC TRUST

YOGA MEDITATION

Caring for your energetic body is a necessary component of health. In my practice, I recommend that women who are struggling with their libido or wanting to become pregnant use the sacral chakra meditations found on page 207.

definitely done my own disconnecting! But what struck me about women's medicine was this need to start bringing them in to being *active participants.*

When a woman is struggling with her libido and having issues with her fertility, I want to pull her awareness back into that space on an energetic level. One way to do that is with a daily meditation, and I highly recommend this even if you aren't having these issues. See the upcoming Brighten Libido-Boosting Protocol for helpful meditations you can try (page 206).

Is the Pill Compromising Your Future Fertility?

Your libido isn't the only part of your sexual reproductive health that the pill can damage. In a 2015 study of 887 Danish women, it was found that women aged nineteen to forty-six who were on the pill had 19 percent lower levels of anti-Müllerian hormone (AMH) and 18 percent less early-stage follicles. AMH indicates a woman's ovarian reserve, so if those levels are low, it can mean she has fewer viable eggs and may struggle with getting pregnant. What was interesting about this study was that even after correcting for all the variables we know can negatively impact these values—age, body mass index, smoking, maternal age at menopause, maternal smoking during pregnancy, and prematurity—AMH was found to be up to 30 percent lower and AFC (antral follicle count) up to 20 percent lower in women using the pill versus non–pill users. On top of that, the pill shrinks the ovaries! The researchers in this study concluded that while the pill makes the ovaries appear "old," this effect may not be permanent. Phew! Also, in case you're tempted, do not test these markers while you're on the pill because they are not an accurate representation of your fertility.

While several studies offer hope by hypothesizing that the changes may not be permanent, just about everyone has stated that women who have taken the pill take longer to conceive, which means it delays fertility. The research has shown that the length of time does have an impact, and the longer you're on the pill, the longer it can take to get pregnant. Even after correcting for age, use of the pill before trying to conceive was found to be associated with a longer window before conception.

However, if you're thirty-five or older, have a history of smoking, have a family history of early menopause, or have any of the other risk factors that compromise fertility, then the pill is only adding a burden to your body and potentially contributing to compromised fertility. In addition, the pill may mask a significant reduction in fertility markers and hormone imbalance, so you may not even realize your fertility is compromised until you decide to come off the pill in preparation for pregnancy.

Women often are told that the pill will have no impact on their fertility. Some are even told they will be more fertile right after they stop taking it! But how can something that causes inflammation, depletes nutrients, disrupts brain communication, shrinks ovaries, and is powerful enough to shut down ovulation not have some level of impact on hormones and fertility? And why don't doctors talk about this? Unfortunately, the emphasis of most of the research pushes promising post-pill fertility outcomes, and like most doctors, I too was taught that the impact of the pill on fertility is negligible, and the benefits outweigh the risks. But the research is contradictory regarding fertility outcomes, and if you're going to take the birth control pill, it's important for you to be informed about the potential risks to your future fertility.

Studies have shown that it can take from several months to over a year for women to become pregnant after taking the pill. The average couple may need six months to conceive, which is why many doctors are unconcerned, even when a woman's period doesn't come back. In my clinical practice I have met with women who struggled to conceive after having had regular cycles before starting the pill, previous pregnancies, and zero indication this would be an

Lab Tests for Fertility

Consider having the following tests to ascertain your fertility:

- anti-Müllerian hormone (AMH)
- DHEA-S
- estradiol
- fasting insulin and fasting glucose
- follicle-stimulating hormone (FSH)
- 4-point salivary or urine cortisol
- luteinizing hormone (LH)
- nutrient testing
- progesterone
- prolactin
- sex hormone–binding globulin (SHBG)
- thyroid panel: TSH, total T4, total T3, free T4, free T3, reverse T3, anti-TPO, and anti-thyroglobulin antibodies

If you are on the pill, you will have an altered AMH, so when you come off the pill if you find you have an abnormal AMH, do the 30-Day Brighten Program and then retest, because I have seen levels improve with these methods.

issue until they came off it. In a small study of women undergoing fertility treatments, it was found that long-term use (ten or more years) of birth control was associated with thinner endometrial lining, the lining of your uterus. The endometrium is where a fertilized egg implants itself, and it needs to be a certain thickness for a baby to want to nestle in. Think of it like a big down comforter on a comfy bed versus sleeping on the floor. Which makes you want to get all snuggly? Animal studies have also shown long-term progestin use is associated with a reduction in estrogen receptors in the endometrium, which means even if estrogen is normal in lab tests, it won't be able to stimulate adequate growth. The result? You get pregnant, but the uterus can't support the baby and a miscarriage occurs. The worst. The mechanism that may

---

Did You Know Your Ovaries Taste?

Yes, your ovaries actually taste the environment and survey for bitter. When bitter is detected, fertility is much more optimal. Many experts believe this provides information about the environment and can have a lot to do with blood sugar regulation. The 30-Day Brighten Program provides you with an optimal diet that's low in sugar and high in healthy nutrients.

---

be leading to fertility issues may very well be how the pill can be protective against endometrial cancer. We need larger studies to understand the impact of the pill on fertility, and really, there is still so much we don't know about women's fertility.

If you're taking the pill and are in your twenties or thirties but want children in the future, you may want to consider the potential impact the pill can have on your cycle and fertility. All the hormonal imbalances that women encounter after the pill can make it extremely difficult to become pregnant. Women with insufficient levels of thyroid hormone also have an increased risk of infertility and miscarriage, and as you know from chapter 7, the pill undermines thyroid health. Leaky gut and autoimmune disease can hinder your fertility too, and as we discussed previously, the pill has been linked to both. The 30-Day Brighten Program can help remedy the ways in which the pill has undermined your fertility. If you know you want to have a baby, start taking steps now to improve your health, and please do not wait until the day you do want to be pregnant to start caring for your fertility.

Progesterone and Pregnancy

By now you know that the pill can affect your progesterone levels, but what you may not know is that this hormone is pretty darn important when it comes to your fertility. Progesterone rises following ovulation and creates an environment in the uterus to support a baby's growth by thickening your uterine lining, reducing inflammation, and changing secretions.

If you struggle with PMS symptoms—mood swings, anxiety, fatigue, low sex drive, or sleep disturbances—you may have low progesterone. I recommend

Can the Pill Cause Melasma?

While most people associate the gray-brown patches of skin on their face with pregnancy, the pill can also trigger melasma. Taking B vitamins and avoiding the sun can help in treating it, but if you use the pill to treat acne, you may just be trading one skin symptom for another.

having your progesterone levels tested if you're trying to become pregnant, since a low amount can make it difficult to get—and remain—pregnant. In addition, adrenal and thyroid function are intimately involved in fertility and progesterone production, which is why these hormones should also be assessed. If you want a healthy libido and a fertile body, make sure you have balanced levels of progesterone.

Optimize Your Health Before Conception

One of the most important messages I want to give you about your fertility is that just because you *can* get pregnant right away doesn't necessarily mean you *want* to get pregnant right away—especially if you have been on the birth control pill. First, after all the nutrient depletions that have occurred in your body while on the pill, you really want to have time to rebuild those nutrient stores so your health and your baby's health can be at their best. Also, if your hormones have been imbalanced, or there are underlying issues that were driving PMS prior to going on the pill or while you were on it, there has been evidence that you may have a more difficult postpartum. If you were having mood symptoms, you especially want to resolve them and find the root cause before you get pregnant, because you'll be at a higher risk of postpartum depression. The pill also causes inflammation, which not only contributes to a greater risk of depression but also is associated with miscarriage. Finally, if you are a woman who was at a higher risk for stroke or blood clots while using hormonal contraceptives, it's important to understand that when you become pregnant that risk will remain elevated—and it could even increase—so this is one more reason to get off the pill, optimize your health before trying to conceive, and check in with your doctor.

A 2018 cohort study of 1.1 million Danish children published in *The Lancet Oncology* revealed an increased incidence of childhood leukemia when mom was taking a combined birth control pill within six months prior to conception. The associated risk was lower when women were off birth control more than six months before becoming pregnant. Look, I want to be clear that this isn't a "shame game" or meant to make you feel bad. We deserve to know how the pill impacts our body and our future baby's body. Today and tomorrow are an opportunity to learn more and do better for our health, and I sincerely hope the research continues so that all women can make the best decision for their bodies.

I recommend that you ditch the pill and start the 30-Day Brighten Program at least six months prior to conception (a year is even better). It takes ninety days to mature an egg, so you need at least three months to optimize the egg before conception. When you are ready to get pregnant, be sure to have all the proper testing through your doctor first to get to the root cause of any issues and restore your health, because mama is both the seed and the soil, which means your focus needs to be beyond just getting pregnant. Growing a human is a pretty big deal, and it takes a lot of necessary resources like nutrients, energy, healthy hormones, and a well-functioning gut, so I believe in a holistic approach to fertility that will establish a lifetime of health for you and your baby. A healthy, happy mother creates a healthy, happy baby. Focusing on your health and wellness before a baby arrives will give you the energy you need to function at your best.

The Brighten Libido-Boosting Protocol

The good news is that there are plenty of natural ways to jump-start your libido and help you achieve those orgasms you've been missing. As you now know, the pill hijacks your testosterone, which is partly what makes your libido plummet, so finding ways to rebalance your hormones and increase your testosterone is crucial. While I know it's a personal choice and therefore not one of the actual steps of the following libido-boosting protocol, I strongly encourage you to consider ditching the pill if you truly want to regain your libido. The research is clear, so if you want that libido to come out of hibernation, quitting the birth control pill is going to give you much better odds. There are so many alternative methods of contraception, which we'll discuss in chapter 13—why stay on the one that gives you a less than satisfactory sex life?

Here are the key steps to boosting your libido—**DATE:**

Decrease stress and increase sleep.

Activate your lady parts.

Talk with your partner.

Eat a libido-loving diet.

Decrease Stress and Increase Sleep

When you're stressed out, your body is in sympathetic overdrive, which means that you're in that fight-or-flight mode during which your body pushes high levels of cortisol at the expense of testosterone, making it especially difficult to achieve orgasm. Make a concerted effort to reduce stress by taking deep breaths, meditating, exercising, eliminating toxic relationships and negative self-talk, and carving out time in your schedule for activities you enjoy.

It's also really important to get adequate sleep when you're struggling with a low libido, since it revitalizes your body and helps rebalance your hormones. When you interrupt your circadian rhythm and get only a few hours of sleep per night, it takes a major toll on your hormone balance. See "Reset Your Circadian Rhythm" (page 233) for tips on getting a good night's rest.

SACRAL CHAKRA MEDITATIONS

Sit at a desk with your feet flat on the floor and your sitz bones at the edge of the chair so that your back is away from the chair and nice and upright, or sit in a cross-legged position on the floor—whatever is the most comfortable for you. Make an inverted triangle with your hands with the heels of your palms on your hip bones, your fingertips pointed down, and your thumbs connected to make the baseline of that triangle. This is the symbol for your womb space and the symbol for women. Bring your awareness into that space and draw your breath down into it. Inhale down into the pelvis, let that breath swirl down, and exhale. As you exhale, release the things that are no longer serving you. Try to do this meditation for three to five minutes every day, and if you can do this twice a day, even better.

If you're a woman struggling with low libido, visualize blood flowing into your pelvis and to those sexual organs, because they need that blood flow and good nutrition, and that's part of arousal and lubrication as well, so just breathe into that space. Hold the color orange in your mind's eye as you do this, or you can drop that color orange right into that pelvic floor space.

Another way to help ramp up your libido (and fertility) is to wear orange panties to light up that sacral chakra.

Ten Foods for Maximizing Your Libido

1. **Dark chocolate:** The bioflavonoids in dark chocolate have a positive effect on your blood vessel health, increasing healthy blood flow to your sexual organs for better arousal, lubrication, and orgasm. Dark chocolate stimulates the release of dopamine, the neurotransmitter in the brain connected to pleasure. Be sure to eat chocolate that is 70 percent or more cacao.

2. **Pumpkin seeds:** These seeds are high in zinc, which can boost your testosterone levels and therefore increase your sex drive. Try to consume at least ¼ cup of pumpkin seeds per day, or try the Flax and Pumpkin Balls on page 301.

3. **Garlic:** I know, I know, this may seem a bit antithetical to getting you in the mood when you're worried about garlic breath. That said, the allicin in garlic thins your blood, improving circulation so those tissues in your lady parts receive more blood and nutrients, potentially enhancing physical sensations.

4. **Oysters:** It's not a myth. These little guys are loaded with zinc, which supports a healthy libido in both women and men.

5. **Pineapple:** This tropical fruit contains bromelain, an enzyme that can increase your testosterone and libido and decrease inflammation. The core has the highest amount of bromelain.

6. **Celery:** Celery contains something called androsterone, a constituent that helps your body produce pheromones. Pheromones

Activate Your Lady Parts

KNOW YOUR LANDSCAPE

You might need to explore your own terrain and understand how your body works. Take your own anatomy tour, and use toys if they help. You may need more lubrication as well, because the birth control pill can lead to vaginal dry-

get secreted through your sweat glands, and they subconsciously suggest sexual arousal.

7. **Ginger:** This anti-inflammatory herb also increases circulation, lubrication, and sexual sensation, thereby enhancing orgasm. On your next date night, swap out that cocktail for some ginger tea and see how much it improves you—and your partner's—sexual performance. Even better, you won't have a nasty hangover in the morning.

8. **Spinach:** This leafy green is an excellent source of the amino acid arginine and folate. Arginine can help the female libido and orgasm through the dilation of clitoral blood vessels. Folate helps aid in the production of histamine, which is released from mast cells during sexual arousal.

9. **Strawberries:** Vitamin C, like what you find in strawberries, supports adrenal gland function, which is where DHEA is made. DHEA can be converted to testosterone.

10. **Avocado:** The good fats found in avocado support a healthy cardiovascular system, which is necessary for tissue engorgement during arousal. They are also rich in potassium and B vitamins that support stamina.

ness. If that's your issue, use a "clean" lube that has no environmental toxins like phthalates. Also consider exercise that gets you into your pelvis, like belly dancing, hula-hooping, hula dancing, and dancing in general. The testosterone-lowering effect of the pill also leads to muscle atrophy in the pelvic floor, making activation of these muscles critical for your health.

APPLY TOPICAL DHEA

As discussed in chapter 2, DHEA, or dehydroepiandrosterone, is a steroid hormone produced in the adrenal glands that is known to improve vaginal lubrication and libido, and it promotes healthy pelvic floor muscles, among other things. You can apply a topical DHEA to the vagina to help with a low libido or if you have pain with intercourse. I do recommend having hormone testing first to determine if you have low DHEA or low testosterone. Remember, it's always important to discover the root cause of your health symptoms.

EXPERIENCE THE MAGIC OF MACA AND GINSENG

Maca, the adaptogenic herb I introduced in chapter 7, supports your ovaries, adrenals, and sex drive. It is a powerful root known for balancing hormones and enhancing sexual desire. (Bonus: It also can help improve sperm motility and erectile dysfunction if your partner is male.) Some people find maca to be a little too stimulating, and it can cause stomach discomfort, so start small and gradually work up to the full daily dose of 1.5 to 3 grams per day. You also can sip it as a beverage or add it to your daily smoothie.

Ginseng is another adaptogenic herb that can help modulate your stress response on a physiological level, which can then help your libido. It has also been known to increase sex drive. In a double-blind, placebo-controlled study of women lacking sexual desire, a supplement containing ginseng—among other herbs, vitamins, and minerals—helped increase participants' sexual desire and satisfaction as well as their frequency of sexual desire and intercourse. If you're struggling with a low libido, try adding ginseng to your daily regimen.

Talk with Your Partner

Part of getting back your libido also includes having conversations with your partner. If you're having issues with your partner, recognize this and go to counseling. Have real discussions with your partner about what works for you and what doesn't work for you. If the root of your libido issues lies in your relationship, then there is no food or supplement that'll fix that.

Eat a Libido-Loving Diet

If you haven't figured it out by now, diet plays a crucial role in your hormones—and that includes your libido! Strive for well-balanced meals with plenty of healthy fats and vegetables to give your body the vitamins and minerals it needs for hormone production. Remember, the pill depletes a lot of nutrients, so you want to be sure to replace them. I recommend eating foods that support circulation and elevate testosterone, such as the "Ten Foods for Maximizing Your Libido" on pages 208 and 209.

BALANCE YOUR ESTROGEN LEVEL

If you are struggling with a low libido, there's a good chance you're dealing with estrogen dominance, which can make you moody, crampy, and bloated—hardly a recipe for feeling sexy! Consume these nutrients to help balance your estrogen (for amounts, see the Brighten Supplement Protocol on page 246):

- **Calcium-D-glucarate** not only supports the rebalancing of estrogen but also supports the liver detoxification pathways that help you to remove excess estrogen from your body.

- **DIM (diindolylmethane)** improves estrogen metabolism, which results in healthier estrogen metabolites and balanced hormones. It does this because it shifts estrogen into the 2-hydroxy pathway.

- **Broccoli seed extract** supports liver detoxification so that harmful toxins, metabolic waste, and hormones are transformed into safer metabolites and readied for excretion.

- **Fiber:** Aim for a variety of fruits and vegetables to reach a minimum of 25 grams of fiber daily.

Key Takeaways:
Boost Your Libido and Fertility

- Orgasms have extraordinary health benefits, like regulating menstrual cycles, relieving cramps, improving fertility, and increasing antiaging hormones to help you live longer.

- The birth control pill can hijack your libido, resulting in a loss of interest in sex, reduced arousal, and difficulty achieving orgasm. Lame.

- Talk to your doctor sooner rather than later if you know you want to become pregnant, because the pill may impact fertility.

- Just because you can get pregnant right away doesn't mean you should. It's important to resolve any nutrient deficiencies, mood issues, hormonal imbalances, or inflammation first. Start the 30-Day Brighten Program at least six months prior to conception.

PART III

TAKE BACK YOUR BODY

CHAPTER 11

PRINCIPLES FOR GETTING STARTED

It's about to go down! But before you start the program in chapter 12, I'm going to give you an overview of how to follow the program and lay out just how to get off that pill if you're ready. **The 30-Day Brighten Program is an individualized plan that will help you get off birth control, eliminate unwanted hormone symptoms, and make friends with your period.**

Depending on what you discovered in the Hormone Quiz at the beginning of the book, you will follow my specific guidelines for diet, lifestyle, and supplements that will help you take back your hormones. And to ensure your success, I've included a meal plan with recipes that are going to make your hormones sing.

This is the program I prescribe to my patients in my Rubus Health clinic, and it is designed for you to begin it right away. Yeah, because you needed to feel better yesterday. Regardless of if and when you plan to stop taking the birth control pill, this program will give you all the essentials to get those hormones and your health rolling in the right direction. If you decide to continue with the pill at least for the time being, you can still achieve considerable progress in your health by following the program. But sadly, you cannot balance estrogen, progesterone, and testosterone until you're off it. If you're currently on the pill and you plan to ditch it, you should definitely get yourself a contraceptive alternative, and chapter 13 can help you choose which method will work best for you.

While my 30-day plan gives you specific tools for detoxing your liver and healing your gut and hormones with foods, supplements, and seed cycling, it's also important to take a holistic perspective and make sure you're incorporating movement and relaxation into your program and embracing your rebel goddess. If you think that who you hang with, how you talk to yourself, and your personal thought patterns don't have an impact on your hormones, it's time to step back and take a good look at how all these are affecting your life. I'll provide some guidance to help you get that house in order too so you can engage in a radical transformation of your health. And I definitely want you to leverage the fifth vital sign—your menstrual cycle—because that is a gold mine of intel on what your body needs. So let's dive in and start taking back your hormones and your body.

How to Follow This Program

When you're feeling tired, bloated, cranky, or you're dealing with less than desirable hormone symptoms, I know the last thing you need is a pile of work thrown on your plate. That's why I'm bringing you resources and tools like nobody's business to make this as simple as possible. You'll find everything you need in chapter 12, and all you gotta do is take it one day at a time. Here's what you'll find:

- meal plans and recipes for the entire 30-day program
- a sample day to help you integrate your diet, lifestyle, and supplement strategy
- a comprehensive supplement guide to individualize your needs
- lifestyle practices for your specific hormone imbalances
- a 14-day liver detox
- seed cycling, moon magic, and other tricks to help you balance your hormones

How to Get Off the Pill

In my experience, when women are done with hormonal birth control, they are *done*. This can mean just jumping off the pill, pulling off the patch, or making a trip to the doctor to remove that IUD. But I want this to be as easy as possible on you, so here's the deal:

First, if you ever used hormonal contraceptives to treat symptoms, the chance of those symptoms returning is pretty big and it's part of post–birth control syndrome. But we're going to leverage those symptoms within this program because they're going to guide you to your root cause and ultimately the best solutions for you.

Second, you have to decide what you're going to use next to not get pregnant. Even if you want to have a baby, as I said previously, you still want to give your body at least six months to recover, and if you can make it twelve months, that just makes my heart soar. (See chapter 13 for a variety of non-hormonal birth control options.) If you're still on hormonal contraception, start using the fertility awareness method to get in tune with your body and track changes in your cycle (again, see chapter 13). But this will not be a reliable birth control alternative until you've been off the pill for a few months. In the meantime, consider a complementary contraceptive until you dial in your natural cycle.

Once you've figured out your backup method for baby prevention, it's important to start looking at your individualized needs. Here are the general guidelines I give women within my practice:

- If you started the pill solely for pregnancy prevention, are coming off it to get pregnant, or had mildly heavy periods, acne, and other bothersome symptoms, you can start the 30-Day Brighten Program and discontinue using the pill when you take the last pill in the pack. There is no need to wean yourself off the pill, but you don't want to stop in the middle of your pack because it can trigger a period sooner than you'd typically have it. And no one likes a double period.

- If horrendous PMS, acne, bleeding, or any other kind of symptoms is what got you on the pill in the first place, I recommend starting the 30-Day Brighten Program now and staying with it at least three months before stopping hormonal birth control. That might sound a little discouraging or confusing, but don't let it be. Here's the deal: In this program, I outline everything you need to take care of your body, and most women will continue that for another three to six months because it takes at least a month or more to shift your menstrual cycle. The good news is that it takes less than a month to start feeling better. If you're a woman who has been diagnosed with endometriosis, PCOS, or another condition that can result in really heavy, painful periods, absent periods, or metabolic disturbances, you have to be really diligent about loving up your body. So if within this program you feel you need extra support, join our community (visit DrBrighten.com/Resources), because this is something a lot of women go through, and sadly, we often struggle on our own.

- If you've already stopped the pill, you can jump right into the program. I help women on both ends of the spectrum, and this program absolutely can help you restore your hormonal health and ditch the weight, acne, mood issues, and PMS.

As you move into this program, remember that natural therapies take time. Be patient with your body. I say this because often women will come to my practice and say, "I tried Vitex [chaste tree berry], and I took it for ten days, and it really didn't do anything for me." Well, Vitex takes an average of three months to take effect. You will be supporting the foundation of your gut, your liver, and your adrenal glands, because once they're healthy and happy, the sex hormones will fall into alignment and it's easier to control inflammation, which is what makes those periods so painful, those moods so difficult, and that weight so sticky. Think about your health like the *Titanic*—a giant ship that is heading in the wrong direction. It's going to take some time to turn that ship around, and you don't want to change course so abruptly that you run into an iceberg and sink the whole thing. Natural therapies work with your body to turn that ship so it's heading in the right direction again. And that can take time. Pharmaceuticals basically strong-arm your body into submission, which is why they can work a lot quicker in some cases. There's a time and a place for pharmaceuticals, and I'm not judging you if you've used them, but I want you to have realistic expectations for how natural therapies work. The most rewarding part about these natural therapies is that you're not suppressing anything, which means that every day you're working toward improving your health, getting one step closer to your goal.

If you've taken the pill or any other hormonal contraceptives for a long period of time, no worries. You're here now, operating with the best knowledge you have, and we're going to keep you moving forward. So just to recap, if you are currently on these hormones, struggle with debilitating symptoms, and have tried to come off the pill before and it was just a wreck, dive right into the 30-Day Brighten Program in chapter 12, and try to stay on it for three months before coming off these hormones. Build up your body and really love it. Feed it with the nourishing recipes starting on page 291. You will not be able to undo everything that has been done, but you can move a little bit closer to health and then come off your hormonal birth control. I also understand if you're not ready to come off these hormones at all. I've worked with women in my practice who say, "Give me six months" or "Give me twelve months." The irony is that nobody ever needs twelve months, because they usually reach a point where they think, *Wait, I feel this good and I could feel better? Okay, forget this noise. I want to get off these things.* But if you need to stay on the pill a little bit longer, use this program to help keep your body as safe and as healthy as possible. If you're pregnant or breastfeeding, please check with your doctor before starting any new program.

Detoxing Your Life

As we discussed in chapter 5, your poor liver has been working overtime, trying to process all that synthetic estrogen from the pill. To get you feeling better fast, we get this party started with a liver detox during the first two weeks of the program. This is the same protocol I use in my clinic, which pairs a detox diet of liver-supporting foods with physician-grade supplements to create some liver-loving synergy. Because the liver is responsible for eliminating the hormones your body no longer needs, balancing your blood sugar, processing nutrients, and supporting your immune system, it's essential that you support it and give it some love.

There are two phases of detoxification by which your liver processes toxins and hormones: phase I and phase II. A lot of detoxes are heavy on phase I, during which your body prepares your metabolites and environmental toxins for phase II detoxification. When phase II detox is not functioning optimally, harmful metabolites produced during phase I can build up in the system, creating unwanted symptoms like headaches, skin irritation, irritability, and fatigue. This is why some people feel awful during a detox—because they support only phase I, which leads to increased toxins (sometimes worse than the original form) that overburden the body. In this program, you'll incorporate foods that support both phase I and phase II of detoxification in order to prevent this.

Sometimes these foods are not enough, and you'll still need more support in that phase II pathway. This is why part of the detox program includes supplements, which will help get that second pathway upregulated so you're never overburdened by phase I detox metabolites. It is truly one of the best ways we can prevent detox side effects. From my experience, women who leverage supplements obtain enhanced benefits and feel better quicker.

When we talk detox, it is as much about helping clear toxins out as it is preventing them from coming in. Taking a holistic approach is the secret sauce to making detox effective and minimizing side effects. You'll be saying buh-bye to those skin-care and beauty products that contain harmful substances. And, no, this doesn't mean you have to walk around without any makeup on (unless that's your jam). It just means you have to start using brands that don't contain a slew of ingredients that can be hating on your hormones.

And I highly recommend you detox from any negative people or bullshit that may be weighing you down and causing unwanted stress. People can be toxic too, and while you're on the 30-Day Brighten Program and trying to rebalance your hormones, surround yourself with friends and family who support you in your goals—not those who don't. You don't need other people's negative energy and drama messing with your mojo. So swipe left on that guy you know

Liver-Loving Foods

Eat 3 cups of a combination of these items weekly or preferably 1 cup a day:

- artichokes
- beets
- broccoli
- brussels sprouts
- burdock or gobo root
- cabbage
- carrots
- cauliflower
- garlic
- grapefruit
- kale
- onions
- turmeric root

is trouble with a capital *T*. Spend time with the people who make your heart happy, who make you laugh and smile, and who bring peace and positivity into your life.

No Is Enough

I understand that when you follow this program there will be people who have an opinion about your diet. There's always someone who has an opinion about your life. But here's the deal: you don't owe them any explanation as to why you're eating, behaving, or doing whatever you think is best for your body, so if you don't want to explain it, just don't. And if you feel like explaining will actually help a girlfriend understand how she can take better steps toward improving her life, then go for it. Otherwise, save your mental and emotional energy and just say, "No, thank you."

Food for Hormone Balance

Let's clear up some common confusion right now: food is information for your body. It's not just fuel or macros or nutrients. The food choices you make provide information to your body about your environment, and remember, if your body thinks the world is a safe place, then you ovulate. Holla! What you put at the end of your fork is some powerful medicine and has major influence over your hormones—for better or for worse.

We're going to kick you off with a whole-foods, anti-inflammatory diet to resolve nutrient depletions and support hormone balance while also eliminating foods that might be messing with your hormones. When I say "whole foods," I'm talking about the stuff that doesn't come in a box or package. In this program, you'll eat plenty of vegetables, protein to help your liver run its mighty detox pathways, and healthy fats, which will keep your blood sugar balanced and supply your body with the building blocks to create hormones. Before you jump in, take a moment to get your mind, home, kitchen, and life ready to rock this detox.

Remember, if you want to balance your hormones, you need to heal your gut too. One of the ways toxins make their way out of your system is through your gut. If your bowels are not moving through this detox process, you're not going to effectively eliminate disruptive hormones and you're not going to feel well. If you get constipated or don't have a bowel movement every single day, then your estrogen will go back into circulation in your body, leading to estrogen dominance. I have designed the 30-day program to include the fiber, the fluids, and everything else you need to get that gut moving. Fiber is excellent for bulking the stool and helping you move out waste as well as cholesterol and fat-soluble toxins. It's also going to help with your hormones and make you feel full longer, which aids in weight loss. Aim for a minimum of 25 grams of fiber per day. You'll do that through smoothies and whole-food meals.

Foods to Eat

There are so many foods to enjoy! Here are some suggestions. Choose vegetables and fruits that are organic, locally grown, and in season, and eat a variety.

Organic Vegetables

Artichokes, asparagus, beet greens, beets, bok choy, broccoli, brussels sprouts, cabbage, carrots, cauliflower, celery, chives, collards, cucumber, endive, green peas, kale, leeks, lettuce, mustard greens, onion, parsley, pumpkin, radish, spinach, sprouts, squash, string beans, sweet potatoes, swiss chard, turnips, watercress, yams, zucchini

Organic Fruits

Apples, apricots, bananas, blackberries, blueberries, cantaloupe, cherries, cranberries, figs, grapes, kiwi, mangos, melons, papaya, peaches, pears, pineapple, plums, pomegranate, raspberries, strawberries

Legumes

Garbanzo beans, kidney beans, lentils, pinto beans, split peas—any variety of legume except peanuts and soy

Seeds and Nuts

Freshly ground flaxseeds, macadamia nuts, pecans, pistachios, pumpkin seeds, sesame seeds, sunflower seeds, walnuts

Reread chapter 6 if you want to remind yourself about the gut, why you need to eliminate certain foods, and pick up some helpful tools. While the 30-Day Brighten Program meal plans are designed to help you restore gut health, some women need additional support in healing. If your gut health quiz in chapter 6 revealed that you're experiencing gut imbalances and symptoms, then it's important to dial in gut-healing protocols that will be vital to restoring your health. See the Brighten Supplement Protocol on page 246 for information on which supplements to add to your diet.

Quality Protein

100 percent grass-fed or pasture-raised beef, bison, buffalo, chicken, eggs, elk, lamb, pork, turkey, venison

Wild-Caught Fish

Cod, croaker, flounder, haddock, mackerel, salmon, sardines, sole (Use seafoodwatch.org to choose quality fish.)

Healthy Fats

Avocados, avocado oil, coconut oil, cold-pressed olive oil, macadamia nut oil, olives

Seasonings

Basil, chives, cilantro, dill, ginger, mint, oregano, parsley, rosemary, sage, tarragon, thyme, turmeric

Beverages

Decaffeinated green tea, herbal teas, natural sparkling seltzer, water

Foods to Remove

Remove these hormone-disrupting foods from your diet.

Gluten and Grains

Barley, kamut, oats, quinoa, rye, spelt, wheat, all gluten-containing products

Dairy

Animal milks, butter, cheese, cottage cheese, cream, ice cream, nondairy creamers, yogurt (Ghee and camel's milk may be okay. Don't have them for at least two weeks, then introduce them and record your symptoms.)

Corn and Corn Products

Popcorn, tortillas, tortilla chips

Soy

All soy and soy-containing products, including edamame, meat substitutes made from soy, tempeh, tofu

Peanuts

Peanut butter or products containing peanuts

Processed/Added Sugars

No agave, corn syrup, high-fructose corn syrup, NutraSweet (aspartame), saccharin, Splenda, white or brown sugar (Stevia, less than 1 teaspoon of honey, and maple syrup are okay.)

Coffee and Other Caffeinated Beverages

Coffee, espresso drinks, caffeinated tea, "energy" drinks, soda

Alcohol

Beer, liquor, wine

Inflammatory Fats

Canola oil, corn oil, cottonseed oil, fast food, margarine, mayonnaise (check the label, because avocado or olive oil base is okay), peanut oil, processed food

Curbing Your Cravings

Okay, let's take a moment to recognize that the pill was depleting nutrients, taxing your adrenals, wrecking your gut, and messing with your hormones in ways that pretty much place any woman at the mercy of cravings. Remember that amino acid tryptophan we talked about in the mood chapter? And that it's necessary for serotonin (which is produced in your gut primarily, BTW)? Tryptophan also helps control appetite, along with dopamine. If you're jonesing for a cupcake, then supplement with 50 to 100 milligrams of 5-HTP (5-hydroxytryptophan) plus its necessary cofactors vitamins B12 and B6. Rhodiola also helps regulate healthy levels of neurotransmitters as well as cortisol (*psst*, that's why I put this in the Adrenal Support supplement). And even though you'll be ditching grains, I want to be crystal clear that you are not going low carb here. Instead, you will be moving grains out and making room for more veggies, which are essential in taking back your body from these hormones and undoing the damage they have caused.

Pay attention to those cravings. If you find they are worse before your period or you are a fiend for sugar, salt, or carbs, then this is a sure sign your hormones need love and that this program is absolutely essential for you. Try incorporating root vegetables, like beets, turnips, and carrots, the week before your period. Low blood sugar is a common culprit for sugar cravings, so make sure you're eating regular meals. Fill up on fiber, fat, and protein and you'll be less inclined to want to grab that office donut (the struggle is real). Recovering your gut and restoring your microbiome will also go a long way toward curbing those cravings. You can do this!

CAFFEINE

You may find it hard to give up caffeine because you love your morning cup of coffee, and hey, I get it. I'm not gonna lie: I love coffee too. I mean, I live in Portland and coffee is like the liquid sunshine of the Pacific Northwest. It's a ritual for most of us, which is why I recommend replacing that cup of joe with something else and taking a break from the bean. If you miss the actual taste of coffee, try one of the alternatives in "Caffeine-Free Coffee Replacements" (page 122). If you could not care less about the taste but rely on coffee for a daily energy boost, start your day with a green juice to get you going and leverage supplements to enhance your natural energy production. Your adrenals and hormones will love you. Plus, you may notice a positive side effect of feeling less stress and anxiety, not to mention sleeping more soundly. No joke. If you're a high-stress gal, you'll be amazed at how different you'll feel after 30 days. Hell, after two weeks. See, I've got your back.

DAIRY

If you're dying for some dairy, you can try coconut, cashew, or almond milk. I find most women have no issue with camel's milk, but go animal milk free for at least two weeks before bringing this in. Swap out dairy cheese for nut cheeses and nutritional yeast. You can also substitute canned, full-fat coconut milk for many cream sauces.

SUGAR

Sugar cravings can be the most challenging, and you may feel hella cranky and tired at first as your body goes through sugar withdrawal. There's literally sugar in *everything* (hello, chicken broth?), so even if you don't devour cookies every night, you're probably getting way more sugar in your food every day than you think. Plus, let's face it: many of us turn to sugar for a quick reward when we're stressed out or down in the dumps, so it's loaded with baggage. If you find yourself struggling with your sweet tooth and tempted to binge on all the foods, do yourself a favor and grab protein first; this will give your body a chance to consider if what you really crave is in your best interest. If you're really missing dessert or feeling left out, try a bowl of fresh berries with chilled coconut cream. That said, if you can avoid giving in to these cravings, you'll have a better chance of ultimately breaking up with your sugar habit for good. Rebound relationships are never a good idea, right?

ALCOHOL

Speaking of rebounds, giving up alcohol also has the benefit of helping you avoid drunk dialing your ex. But I get that it's a major part of socializing and you don't want to feel like Debbie Downer at a party or go into hibernation mode for 30 days. But who says you need alcohol to have a good time? Order a fun mocktail when you go out with friends or try one of the recipes in the smoothies (page 291) or beverages (page 304) sections on a night in with your significant other or lady tribe. Serve it up in a wine or champagne glass to put yourself in a celebratory mood.

What to Expect

As you begin the program, detoxing your liver and healing your gut with the right foods and supplements, you may encounter a few, shall we say, unpleasant side effects. This is totally normal. You're making some big and important shifts in your body, and there will be a transition as you come off synthetic hormones. As I previously mentioned, you may feel a bit cranky and tired at first. You may even have a day when all you want to do is nap on the couch in your yoga pants. If you're cutting out caffeine and sugar cold turkey, there's a good chance you'll experience headaches for a day or two as your body adjusts. Imma be real with you here: it's not all sunshine, rainbows, and unicorns for some of us when we kick these hormones. And if you find you struggle in the beginning, then know you're not alone and you can move through this to enjoy better health. Check out our community at DrBrighten.com/Resources.

But don't be discouraged. As you support your natural detox, mend your leaky gut, and balance your hormones, you'll gradually notice much more appealing side effects. You'll experience bangin' concentration and energy during the day and sleep soundly through the night all month long. Your mood will improve, and feelings of anxiety or depression that were the result of the pill or PBCS may begin to diminish. Over time, your skin will clear as well, though skin is typically one of the last organs to heal so be patient with this one. And if you do have a flare-up, use the guide on page 53 in chapter 3. As you support balancing your hormones, you'll also become aware of changes to your menstrual cycle. That missing period may finally return. Or it may get lighter and less painful. And if you have late, irregular, or short cycles, they may start to even out and become more regular. Even better? All those PMS symptoms of cramps, bloating, mood swings, and headaches will improve. You'll no longer suffer from migraines. Another bonus? Healing your gut may help you get rid of a gut if you have one, because belly fat is often the result of inflammation and you'll be getting the inflammation in your body way down. Plus, there may just be a little more bow-chicka-wow-wow going down when your libido decides to make its debut.

Preparing for the Program

Before you start Day 1 of the program, read through the next chapter to understand what individualized practices and supplements you'll need to get your hormones back in alignment. Use the supplement table on page 246 along with the assessment results on pages 20–21 to create your custom plan of success. Clear out of your pantry and refrigerator all the "Foods to Remove" listed on page 224, and make a trip to your local grocery to restock with foods that can't help but love your body up. Cue Marvin Gaye.

It's also helpful to prepare some meals ahead of time, because if you're having a busy day, you can't just fall back on your usual takeout habits. And when you're tired and hungry—or hangry!—it can be hard to stick to a plan, unless you've prepared food ahead of time and always have go-to snacks on hand. I've tried to make the meals as simple yet flavorful as possible, because I know you have a busy life and don't necessarily want to spend all your time in the kitchen.

Make It Move, Baby!

While you're on the 30-Day Brighten Program, exercise, sweat, and bust a move! Activity will get your lymphatic system moving and your blood flowing, and the sweat will eliminate waste from your body. Exercise also increases lean muscle mass, which helps prevent osteoporosis and increases insulin sensitization. Studies suggest that improving insulin sensitivity can dramatically improve your hormones, not to mention reduce your risk of diabetes and heart disease. Winning! Aim for thirty minutes of exercise five days each week to support thyroid hormone conversion, drop stress hormones, and improve circulation. Plus, exercise gets your bowels moving, so you can rid yourself of excess estrogen!

You can exercise for thirty minutes all at once or break it up into three ten-minute sessions within the day. If you're having a low-energy day, it's a good time for activities like restorative or yin yoga, stretching, light swimming, or light Pilates—or just get out and do some walking or gentle dancing. If your energy levels are strong, fit in some weight lifting, high-intensity interval training (HIIT), more rigorous yoga or Pilates, jogging, swimming, or any other sport or activity you enjoy.

Kick Stress to the Curb

Just a reminder that stress does your hormones no favors, so make sure you're doing your mind–body work. As I've said throughout the book, stress wreaks havoc on your hormones, so this is a critical component of the program. Let that shit go, girl. Any time you do a detox, you also want to support your emotions and your stress, making sure you're detoxing those as well.

Embrace Your Rebel Goddess

If you haven't gotten the memo, you're a badass. You have the power to create and birth an entire being into this world. You are part of the tribe—the woman tribe—that has ensured the existence of the human race. And for as far back as our history goes, we women have been either cherished and honored or shamed and made outcasts for the undeniable power we possess. Even now, as you read this, you may be feeling a *Hell yeah* or a *Get me the hell out of here* bubble up. Honor it. Sit with it. And know that this has more to do with what you have lived through before and outside influence than with your inability to connect to your power source.

Whoa! Did we just make a sharp turn into "Woo, woo"? Call it what you want, but, yeah, we just did, because I'm not holding anything back in this book. Let's take a reality check and understand that your physical body is only one of the many layers of you. And if you're serious about healing, serious about reclaiming your body and your power, then you've gotta tap into your inner wisdom. Now take that rebel goddess fist and raise it to the sky. Because you are a powerful goddess, and anyone who questions that or makes you feel you are the lesser can jog on.

Your Cosmic Rhythm

Have you ever noticed that some of your friends get their periods around the new moon and some with the full? That's because we ladies are in sync with the cosmic rhythm of the universe. All living creatures are influenced by light–dark cycles, like what you get with the moon. But we ladies are extra special because our periods and ovulation are synced with the moon. (*Psst*, some of us also have waxing and waning moon cycles, which is okay too.) This is something many women have observed, and it was reported in primates long before science introduced the pill. Louise Lacey—who recognized in the 1960s that the pill was blocking her natural rhythm and that when she stopped, her periods became irregular (hello, post–birth control syndrome)—took to charting, studying the moon, and cultivating her circadian rhythm, and in 1975 she published about them in her book *Lunaception*. Back in 1978 Dr. Edmond M. Dewan demonstrated that exposure to light can affect ovulation and be used to regulate a menstrual cycle. Back in 1978! Back before we were inundated with LED street lamps, iPads, cell phones, and constant light exposure. Of course good ol' Ed was criticized and his findings were mostly ignored, but we have come to understand that he was seriously on to something. In addition to messing with your cycle, artificial light has been found in multiple studies to have correlations with elevated breast cancer risk in women who work at night.

In more startling news, in 2001 the *Monthly Notices of the Royal Astronomical Society* reported that "two-thirds of the U.S. population and more than one-half of the European population have already lost the ability to see the Milky Way with the naked eye." Let that sink in, especially if you're a gal with irregular cycles, a family history of breast cancer, or an increased risk of developing breast cancer (*cough!* the pill). Artificial light exposure at night affects your menstrual cycle, can influence ovulation, and is correlated with an increased risk of breast cancer . . . and we're living in the most light polluted time in history—so much so that many of us can't even see any stars from where we live. This is a big problem, so obviously I'm gonna give you some solutions to maintain your cosmic connection as a modern-day woman. I will teach you how to do this plus leverage food, supplements, seeds, and herbs to help you get your cycle in sync in the next chapter.

Let's start by breaking down some moon knowledge. If you get your period on the full moon, that's considered a red moon cycle, whereas bleeding on the new moon is a white moon cycle and is thought to be more common. Are you a white moon goddess or a red moon goddess?

Red Moon Cycle: Period on the Full Moon, Ovulate on the New Moon

Have you heard you "should" menstruate with the new moon and ovulate with the full moon? Yeah, I was told that too in medical school, which propelled me into a deep dive into the old-school textbooks, like the kind that just might fall apart if you're not gentle enough (I was lucky enough to have a rare book room at my school). What I found was that based on a woman's greater purpose—what she's currently feeding with her energy and attention—her period will fall on a full or new moon. The red moon goddess is often a medicine woman, shaman, healer, or priestess; in other words, she's here to do some big work. Translated to modern day, these are doctors, nurses, CEOs, activists, artists, and all-around movers and shakers. These women tend to focus their period powers and energy outward into the world during their period. Turns out, that "should" isn't right, and in my opinion it's a bit silly to think the only thing we sync with in the universe is the single moon we see. So don't get hung up on shoulds. In fact, F that noise and get in sync with what is true for you.

White Moon Cycle: Period on the New Moon, Ovulate on the Full Moon

Do you ovulate on the full moon? This is thought to coincide with a fertile cycle (aka ready to make that baby). Think about it: being in a cave, the full moon waking you up. What else are you going to do? When light enters your eyes and hits the pineal gland in your brain, it triggers your body to degrade melatonin and you wake up. This is why some women have trouble sleeping leading up to and during a full moon. All that extra light shining down is also why the Earth is considered most fertile during the full moon. But enter light pollution and we've got a whole other ball game when it comes to messing with your hormones. Remember, melatonin is also an antioxidant, and your ovaries have receptors for this hormone.

The white moon goddess tends to pull her energy inward with her period as a way to nourish herself and tap into her intuition. She's running energy to create life, which is why this is thought to be the cycle of a woman whose body is ready for a baby. But don't be fooled into thinking she's just here to breed. The white moon goddess has some seriously deep knowledge and intuition, and what she has cultivated might just blow you away when she gifts it to the world.

When Your Moon Cycle Shifts

You can switch your moon cycle. In my practice, I've seen women's cycles begin to shift to help them sync up to ovulating with a full moon when they've got baby making on the agenda. I've also seen it happen in women who are changing where they focus their attention. This does not mean you can't get pregnant if you ovulate with the new moon. You definitely can. There are plenty of ladies in my clinic who have become pregnant with their period falling on the full moon. I personally was a red moon goddess when I conceived and following the birth of my son. I had nearly five years of consistent red moon cycle until I accepted a partnership with my publisher and decided to birth this book. My cycle days seemed to be slightly off each month, but my intuition and my observation of thousands of women told me there was wisdom in this. Writing a book requires being withdrawn (ask my friends), turning your attention inward, and cultivating some serious knowledge. So shifting to a white moon cycle made sense, and I gotta say it's been a radical ride to be ovulating with supermoons. I'm excited to see how my cycle shifts as I move into the next phase of speaking about this book to the world. I'll keep you posted in our community at DrBrighten.com /Resources.

What if your period is coming several days before or after the moons? Well, it may be you're shifting cycles, or you're being governed by forces beyond the moon. When all you see is the one moon from our planet, it's easy to think that's all there is, the only piece connecting you to the cosmos. But in reality, you're a divine being who is connected to the cosmic rhythm of both the seen and the unseen.

Cosmic Cycle Sync

To maintain your cosmic connection and sync your cycle, you need to get out into nature and expose yourself to sunshine and fresh air and do some "earthing." This means connecting to the earth by placing your bare feet on the ground, lying in the grass, or hugging a tree. In Japan, this is called nature bathing, and it's a serious stress reducer that will help you and your body recognize that it's daytime.

Also helpful is charting your menstrual cycle along with the moon phases.

Next, try "pulling down the moon": Go outside on the full moon. If you can stand barefoot on the ground, even better, because it will connect you to the earth. Make a triangle with your hands by placing your index fingers and thumbs together. Extend your arms up to the sky, and place that moon right in the center of the triangle. Visualize pulling down the moon's energy into your sacral chakra, bringing that energy into your pelvis and filling up that pelvic space with nourishing yin energy. (This is also a great practice if you're trying to become pregnant.)

If you're digging all this moon and energetic speak, then I highly recommend checking out *Code Red* by Lisa Lister as a resource. Plus, I just love her.

Reset Your Circadian Rhythm

First and foremost, if you're not sleeping, then you stand zero chance of getting your hormones back on track. It's also an important time for your melatonin and growth hormone to do their jobs. Follow these steps to reset your circadian rhythm:

1. Live by candlelight after 8 p.m. once weekly to promote a change in your circadian rhythm.

2. Wear amber glasses consistently two hours before bedtime and any time you're looking at an electronic screen (TV, computer, smartphone, etc.)—these glasses block blue light to allow the rise of melatonin.

3. Create relaxing bedtime rituals, such as sipping herbal tea, taking a bath, doing some light reading, stretching, or meditating.

4. Take 2 dropperfuls of ashwagandha or 50 to 150 milligrams of phosphatidylserine before bedtime to help decrease cortisol levels.

5. Sleep in a completely dark room, and avoid all light-emitting electronics (TV, computer, cell phone) for those two hours before bed. No more scrolling through Instagram or texting late at night.

6. Keep your room temperature at or below 70 degrees Fahrenheit to optimize sleep.

7. Aim for eight hours of sleep every night, or more, and get in bed before 10 p.m.

8. Expose yourself to natural light upon waking by opening up the curtains or going outside, though the light doesn't have to be direct sunlight. No, "screenrise" (checking your phone when you wake) does not count.

You can also boost your melatonin naturally by taking 150 to 300 milligrams of magnesium nightly and eating pineapple, cherries, bananas, or oranges, which are natural sources of melatonin.

Charting and Tracking Your Menstrual Cycle

Remember how your period is your fifth vital sign? Well, you can be your own health detective by using your menstrual cycle to navigate your symptoms and understand what you need. It's important to have a baseline of where you are before you begin this program so that you can track your personal data and find out what is and isn't working. Keep a record of your symptoms and how you're feeling, and note what improves.

You Got This!

Are you ready to dive into the 30-day program? You can totally do this, and if you ever feel like you need additional support, we have a community you can join. I have an online support program, plus you can always work with my health coaches or my clinic.

CHAPTER 12

THE 30-DAY
BRIGHTEN PROGRAM

Let's do this! You've learned everything, you have all the tools, and now it's time to put it all into practice and take back control of your hormones. This program is going to help you if you're on the pill, if you're coming off the pill, or if you've already ditched it. You're taking the first steps toward increasing your energy, elevating your mood, and eliminating unwanted hormone symptoms. By leveraging naturopathic and functional medicine, clinical nutrition, and mind–body medicine, you can effectively shift your hormones into a state of bliss. This is the foundation of what I practice in my clinic, and I'm excited to share it with you in the 30-Day Brighten Program. Let's begin your healing journey!

Take a look at the Hormone Quiz you filled out in chapter 1 (pages 16–21). What did you learn? Do you have estrogen dominance? Low progesterone? High cortisol? Low testosterone? This can help you start to identify problems and what you need to do during this 30-day plan to hit the reset button. Remember, one hormone imbalance often leads to another, so don't be surprised if you fall into multiple categories. Regardless of which hormone imbalances you have, follow the liver-detoxing, hormone-repairing meal plans (page 259), because those will support your body as a whole and be the fastest way to reset after hormonal birth control.

The Brighten Supplement Protocol table will give you instructions for added support in the particular areas where you need it, based on your individual imbalances. Think of it as a "choose your own adventure" based on your results where you get to zero in on particular hormones. Start taking these supplements immediately, because natural therapies can take time. You'll work with your body to reset its natural rhythms and teach or remind it of how things used to be—and how lovely they can be again. Most women following this program need to remain on the supplements for at least three months. Continue to track your symptoms and use the quizzes and checklists in this book to reevaluate your diet, lifestyle, and supplement needs.

What Your Body Is Telling You

Before you get started on the program, I want you to get your baseline with the Hormone Quiz if you haven't already. This will also allow you to track your progress. Sometimes women in my clinic tell me they have a whole laundry list of symptoms and then three or six months down the line, they've forgotten they even had them. It's a real thing: you can totally forget your periods were ever awful! And I want to help you do that. Most women who are not cycling and go through this program need to stick with it for about three to six months.

Take the Hormone Quiz any time you need a check-in—if you're thinking, *I'm just not sure what these hormone symptoms are telling me,* use the quiz, because it can help you get clarity on what's actually going on. While too much estrogen may be an issue for you as you enter into this program, it's entirely possible that a year from now you'll not have enough estrogen. Any number of factors can affect your hormone balance, so it's good to tune in regularly to see if your hormones need a little help. I often have my patients evaluate their symptoms once a quarter to see how they're doing. If you find that you start checking more boxes, don't wait. Intervene and get a handle on your symptoms right away. Then follow the lifestyle strategies for each category of the quiz and begin with the corresponding supplements in the table.

If you checked fewer than two boxes in any category of the Hormone Quiz, as I mentioned, it's unlikely that this category is a culprit for you. If you checked two or more, then this is an area that needs your attention. And if you checked five or more, then this just might be your troublemaker. It is likely your dominant hormone right now, aggravating your symptoms, and it needs to be balanced.

Your symptoms can guide you. Start tracking them in your menstrual cycle. The first day you see blood is day 1 of your cycle. Know when you bleed, how long, and the length of your cycle (day 1 to the next day 1), plus track when and where your symptoms appear, so you can prepare and start correcting imbalances. For instance, if you figure out that you have anxiety five days before your period and you're going to have trouble sleeping, you can bring on support seven to ten days before your period to head it off at the pass.

Category A: Too Much Estrogen

Too much estrogen—estrogen dominance—is one of the most common hormone imbalances that I see in my practice and in women coming off the pill. It creates a burden over time on the liver, making it difficult to move that estrogen out. If you checked five or more boxes in category A of the quiz, you definitely want to consider the advanced support and supplement recommendations, because often when you've got this much estrogen going on it's rough—you're irritable, you're angry, you want to cry all the time, you have brain fog, you gain weight in your hips, butt, and thighs—and it's not a fun place to be. Also, it can be more difficult to do the things you know are best for you and will help you feel better. I find that women who leverage diet and lifestyle, along with supplements, are most successful in balancing their hormones and get back to feeling like themselves sooner rather than later.

Estrogen dominance can be either frank or relative. Frank estrogen dominance is usually due to environmental toxins or poor detoxification through the liver and the bowels. If you found yourself checking boxes in category C of the quiz as well, which indicates too little progesterone, then it's likely you have relative estrogen dominance: too little progesterone to oppose your estrogen. This is a common cause of PMS symptoms.

If you're experiencing symptoms of estrogen dominance, whether relative or frank, add the supplements from the table on page 249 into your routine during the program and follow these diet and lifestyle practices:

- Avoid xenoestrogens big time—drink only out of glass bottles and/or with metal straws, and use glass containers. Be mindful of takeout food. BPA-free is BS. Don't trust it.

- Eliminate alcohol completely—one alcoholic drink can jump that estrogen up by over 10 percent. That doesn't mean alcohol has to be gone forever, but for the duration of this program, I highly recommend avoiding it.

- Leverage the guides for detoxing your skin care and your house (see Resources). Visit the Environmental Working Group's website (EWG.org/skindeep), and check all your personal care items to make sure you're purchasing the cleanest products on the market. (See chapter 5 for more details.)

- Take a supplement that aids in the metabolism of estrogen and is protective like DIM, chrysin, and calcium-D-glucarate. Consider Balance by Dr. Brighten, which contains these nutrients, plus black cohosh and Vitex for optimal hormone balance.

- Follow the circadian rhythm exercises and consider melatonin—0.5–3 milligrams nightly. Melatonin is not just for sleep but is also a great antioxidant. Low melatonin has been associated with breast cancer risk.

Category B: Too Little Estrogen

This is also common after stopping the birth control pill. Say what? In reality, all hormonal chaos is possible after stopping the pill or any other hormonal contraception! Estrogen naturally drops for us starting in our forties and fifties. With too little estrogen, your skin gets thin, your breasts droop, and you probably don't have much of a libido. The big concern here is that your bones, brain, and heart not age faster than they should.

Too little estrogen may be a sign you need more dietary fat. Keep in mind that dietary fat increases your estrogen in a healthy way. Don't misunderstand that if you eat fat you'll have estrogen dominance. If you eat the wrong kind of fat, such as canola oil or hydrogenated fats, you'll have some trouble, but anti-inflammatory fats are essential to all your hormones.

Low estrogen may also be a sign of menopause. If you're older than forty-five, this may indicate you're heading into menopause, which we call perimenopause. But if you're younger than forty, this could be primary ovarian insufficiency (POI). The majority of POI cases are a mystery, but clinically I have seen that an autoimmune issue is often at the root. If your biological clock is ticking, you can't change that. (I would totally do that for you if I had a magic wand.) You can't change what nature is intending, but this program will support you in transitioning into menopause, because some of the same secret sauce that helps you transition to menopause also helps you get off these hormones and keep your sanity.

If you've recently given birth, too little estrogen is super normal postpartum. You birthed a human, then a placenta, and at that moment your hormones dropped to almost zero, like a postmenopausal woman. So if you're newly postpartum, expect your estrogen to be low. You can use some of the same guidelines in the program to support the return of your hormones and manage symptoms like vaginal dryness. But, of course, always check with your doctor before starting supplements if you're breastfeeding.

If you have too little estrogen, add the supplements from the table on page 250 into your routine during the program and follow these diet and lifestyle practices:

- Increase your dietary fat; it will give you the building blocks so you can actually make hormones. Add more avocado oil or avocado, cold-pressed olive oil, sardines, and mackerel to your diet.

- Consider testing for primary ovarian insufficiency (POI). Never jump to the conclusion that your ovaries are failing; I have seen labs of women whose doctors diagnosed them with infertility turn around with the protocols in this program.

- Use bioidentical hormonal therapy if you're having uncomfortable vaginal dryness or debilitating symptoms, which you can talk to a doctor about. You can also use vitamin E 400 IU as a suppository, which is quite lovely for the health of your vagina.

- As well as trying a supplement of maca, you might add maca to smoothies or other beverages.

- Increase your flaxseed consumption.

- Have regular orgasms. These not only help your hormones come into balance but lower inflammation, improve immune function, and reduce PMS as well.

- Consider removing gluten from your diet, which you'll do in the 30-day program. A low estrogen level may be due to a food sensitivity.

- Take vitamin E and magnesium; they are both excellent for your hormones. They can be helpful if you're having night sweats or hot flashes. Nuts are a great food source, and you'll be consuming hormone-supporting foods in the program.

Category C: Too Little Progesterone

When you have too little progesterone, your mood can be all over the place the week before your period, especially if you've got a ton of estrogen. You may be losing your hair, feeling really irritable and cranky, and looking swollen and puffy. Progesterone stimulates the GABA receptor in the brain and helps you feel really chilled out and calm, so if you don't have enough, you can have trouble sleeping and feel anxious. It's common after age thirty-five to see progesterone decline because we ovulate less frequently. As I mentioned before, low progesterone and estrogen dominance tend to go hand in hand.

With too little progesterone, you have to get your stress down, big time. As long as you're hanging in high stress, your body will struggle to make adequate progesterone.

If you have low progesterone, add the supplements from the table on page 251 into your routine during the program and follow these diet and lifestyle practices:

- Dial down stress. Consider adrenal supportive practices like eating in a relaxed environment, meditating, practicing yoga, exercising regularly, and employing deep breathing. The number one place I start with women is the adrenal glands, because they are central to so many hormone issues.

- Take a supplement containing Vitex and B6 to support progesterone production. Consider Balance by Dr. Brighten, which also contains nutrients and herbs to support estrogen balance.

- Vitamin C as a supplement or in foods can help improve progesterone levels and is a nutrient depleted by the pill.

- You may also require bioidentical progesterone. Discuss this with your doctor.

Category D: Too Much Testosterone

With elevated testosterone, especially if you have insulin or blood glucose issues, get evaluated by your doctor to see if you have PCOS. Yes, your hormones will be cray, but you're also at really high risk for heart disease, stroke, and diabetes if you have PCOS—all the things that hormonal contraception can cause anyway. In the meantime, you can still follow the 30-day program to support your hormones.

Blood sugar balance is at the crux of everything hormonal, but particularly when you have too much testosterone. Blood sugar dysregulation can cause the ovaries to kick out more testosterone, causing symptoms of acne, hair growth on your chin, chest, and abdomen, or hair loss on your head. And nobody likes any of that. You also need to examine your thyroid and adrenal glands.

Too much testosterone may also be an androgen rebound. When you stop the pill, you stop suppressing testosterone, and then boom, now you have *all* the testosterone. You may develop PCOS-like symptoms but not actually have PCOS. You can fix this, but it can be a big reason why women experience so many complications with their skin when they come off the pill.

If you have too much testosterone, add the supplements from the table on page 252 into your routine during the program and follow these diet and lifestyle practices:

- When you have your period, on day 3 test your FSH and LH to rule out PCOS. Also get your insulin and inflammation checked, and get a cardiometabolic panel, a number of tests used to assess cardiovascular and metabolic disease risk. Ask your doctor to check your androgens as well.

- Replace your coffee with some decaffeinated green tea, which can improve sex hormone–binding globulin (SHBG) and testosterone levels. Consider Balance by Dr. Brighten, which contains green tea, plus herbs and nutrients that can balance estrogen and progesterone as well.

- Freshly ground flaxseeds can raise sex hormone–binding globulin (SHBG), and that will help bind up your testosterone so you won't have as much of it bioavailable to your cells.

- Take a supplement with saw palmetto, zinc, vitamin D, and nettle root. These herbs and nutrients help with excess testosterone levels and prevent it from converting into DHT, which causes hair loss.

- Support those adrenal glands. In women with excess testosterone, licorice has been shown to be helpful to bring down those hormones, and it's great at keeping your cortisol around. Consider Adrenal Support by Dr. Brighten, which contains licorice.

- Have some turmeric, because it's going to drop the inflammation and makes your cells more sensitive to your hormones in a good way, but it's also going to help your liver detox and support your ovaries. Consider Turmeric Boost by Dr. Brighten.

- Take chromium, 300 to 1,000 milligrams daily, to help with blood sugar sensitization.

- Take 2 grams of inositol once or twice daily.

Category E: Too Little Testosterone

If your testosterone is low, you have a seriously diminished sex drive, and you may feel depressed, have mood swings, or cry easily, which certainly isn't going to help that libido. Or perhaps you're experiencing anxiety and panic attacks. You're tired or fatigued and seem to have completely lost any and all motivation. You're also losing muscle mass and gaining weight, and you may even have cardiovascular symptoms or heart disease. When you have low testosterone, you often don't feel like your usual self and everything is a struggle.

To remedy a lack of testosterone, add the supplements from the table on page 253 into your routine during the program and follow these diet and lifestyle practices:

- Pump it! Try strength training, Pilates, body-weight resistance exercises, or high-intensity interval training. Exercise builds skeletal muscle and helps sensitize you to estrogen. It stimulates the growth of your muscles, which helps testosterone. Exercise will also lower body fat, and body fat makes estrogen, which pushes you to convert all your hormones into estrogen, including testosterone.

- As women, our testosterone comes from our ovaries and our adrenal glands, so loving your adrenals is a must, as is good blood sugar regulation. Try Dr. Brighten's Adrenal Support.

- Eat an antioxidant-rich diet to protect your ovaries, and consume healthy fats. Incorporate more avocado, spinach, celery, garlic, ginger, pineapple, strawberries, dark chocolate, red meat, and plenty of healthy fats to support testosterone production.

- Reduce stress. Practice some of the mind–body exercises recommended in the program, because as long as you're stressed out, your body pushes everything into cortisol production at the expense of what drives your libido.

- Take vitamin D. Vitamin D deficiencies are associated with low testosterone, so having adequate vitamin D is important for testosterone production.

- Take adaptogenic herbs, such as ginseng.

- Get quality sleep.

- Have sex. You need to be having orgasms. It's a use it or lose it situation.

Category F: Too Little Cortisol

Too little cortisol is a sign of HPA dysregulation, or "adrenal fatigue." Getting adaptogenic herbs and nutrients to feed those adrenals is a must. Three more things your body absolutely needs you to do are reduce your stress, balance your blood sugar, and be super gentle with yourself. That means no going crazy at the gym, trying to lose weight—it will do the opposite of what you think it will with your hormones—and not hating on yourself for not being able to "push as hard" as you used to.

It probably won't be enough to focus just on diet and lifestyle if you have too little cortisol, especially if you're feeling pretty depleted, so make sure to add the supplements from the table on page 254 into your routine during the program. But here is the most important thing you can do:

- Take a good B complex vitamin. You will need B5 for your adrenal glands, B6 for your progesterone and estrogen, and B1 if you have higher inflammation or really painful menstrual cramps. Consider B-Active Plus by Dr. Brighten, which contains active B vitamins, including those that benefit people with *MTHFR* variations.

- Take adaptogenic herbs like Rhodiola, *Eleutherococcus*, licorice, and ginseng, which help support healthy cortisol levels. Consider Adrenal Support by Dr. Brighten, as it contains adaptogenic herbs and key vitamins to help your adrenals thrive.

- Take vitamin C, which is tremendously helpful in all of your hormones and is depleted by the pill.

- Be gentle and kind to yourself, because this actually lowers inflammation. If you catch yourself running unkind thoughts about yourself, slow your roll and reframe. You're amazing. I know it. You know it. So own it.

Category G: Too Much Cortisol

If your life is crazy stressful and overwhelming or you're storing belly fat or you're waking up in the middle of the night with your mind racing—worrying about all the things you need to accomplish the next day or obsessing about everything you didn't get done that day—then you know how it feels to have too much cortisol.

Beyond evaluating the relationships in your life and just who and what is stressing you out, you have to balance your blood sugar and restore your adrenal and thyroid health; otherwise, your estrogen and progesterone are never going to fall into line, and your testosterone is going to do whatever it wants.

If you have too much cortisol, add the supplements from the table on page 255 into your routine during the program and follow these diet and lifestyle practices:

- Eliminate any stressful relationships. Maybe they go forever and maybe they don't, but right now you have to take care of yourself.
- Consider taking adaptogens like ashwagandha, Rhodiola, and Gotu Kola. You can also use licorice, which is anti-inflammatory, but don't take it if you have high blood pressure. Take Adrenal Support to help balance cortisol and replenish your nutrient stores.
- Feeling wired and tired at bedtime? Take Adrenal Calm if you have trouble sleeping or are waking up at 3:00 a.m. and can't get back to sleep—or if you wake up in the morning feeling anxious. Yeah, this supplement is your jam. Adrenal Calm contains phosphatidylserine, an amino acid that helps naturally bring down cortisol, and nervines— herbs that will help you get into the "rest and digest" and out of the "fight or flight" state. It's a two-for-one in supporting adrenal and nervous system health.
- In times of high stress our body requires more B vitamins. Ya know, the ones the pill has been depleting. Consider B-Active by Dr. Brighten. In my clinical experience, women do best when using a B complex in addition to the herbal supplementation.

- Take omega-3 fatty acids. These are anti-inflammatory, support skin health, and help your cells use your hormones.
- Find exercises that help you chill. Yoga (no, you cannot skip savasana), Pilates, walking, and other relaxing forms of movement can make a real difference.
- Meditation, prayer, float tanks, acupuncture, massage, and even getting a mani-pedi can help you chill and master your stress response.

Category H: Too Little Thyroid Hormone

If you checked the boxes for category H, you have to get a complete thyroid panel, because you may need a thyroid medication. Your body needs this thyroid hormone; it's nonnegotiable. Too little thyroid hormone can place you at a high risk for things like congestive heart failure in the long term, and it strains all your other hormones in the short term, making you feel awful. We're talking trouble with your mood, your metabolism, and your menses.

If you have too little thyroid hormone, add the supplements from the table on page 256 into your routine during the program and follow these diet and lifestyle practices:

- As I said, get a full thyroid panel (see chapter 7 for more information).
- If your labs and body indicate you need medication, take the medication. You may not need this medication forever and there may be a root cause that can be addressed, but you have to support your body in the best way you can so that you can one-eighty that business and get back to a healthy life. In my clinical experience, most women do best on a natural desiccated thyroid hormone, like Armour, Nature Throid, or WP Thyroid.
- Start taking nutrients like iron, iodine, zinc, B vitamins, and vitamin C, which are essential to thyroid hormone production. Consider Thyroid Support by Dr. Brighten, which delivers these nutrients, plus helpful herbs to optimize thyroid health.
- Take a prenatal or multivitamin. Most menstruating females do better on a prenatal because they're losing iron during their period, and when you're coming off of these hormones, those levels can often be low. Consider Prenatal Plus by Dr. Brighten, which has a balance of selenium and iodine, along with all those nutrients the pill was depleting, to help you rebuild stores and optimize thyroid function.

Brighten Supplement Protocol*

FOR EVERYONE

Supplement Type	How to Take	Recommended Brands	Notes
Multivitamin or prenatal vitamin	2 capsules twice daily 3 capsules twice daily	Dr. Brighten Women's Twice Daily Multi, Dr. Brighten Prenatal Plus, Seeking Health Prenatal, or Innate Baby and Me	Choose a supplement with activated B vitamins such as methylfolate, B12 (methylcobalamin), and B6 (riboflavin 5'-phosphate). Avoid folic acid.
Lactobacillus and *Bifidobacterium* probiotic	50 billion CFUs daily After the 30-day program, 15–20 billion CFUs daily	Ther-Biotic Complete or Ther-Biotic Factor 4 by Klaire Labs	If you suspect that you have SIBO, skip the *Lactobacillus* and instead begin with *Saccharomyces boulardii* and a spore-forming probiotic.
Saccharomyces boulardii probiotic	500–2,000 milligrams daily	FloraMyces by Designs for Health	Beneficial yeast
Spore-based probiotic	½ capsule for 7 days 1 capsule for 14 days 1 capsule twice daily for 14 days 2 capsules twice daily for at least 60 days	MegaSporeBiotic by Microbiome Labs or Dr. Brighten Mega-Spore	Start with a low dose and gradually increase to avoid digestive upset.

*I'm always staying up to date on the latest research and adjusting my formulations, so please visit DrBrighten.com/Resources for regular updates.

Brighten Supplement Protocol

FOR EVERYONE

Supplement Type	How to Take	Recommended Brands	Notes
Omega-3	2 capsules once or twice daily	Dr. Brighten Omega Plus or Nordic Naturals ProEPA Xtra	Look for brands that use third-party testing, sustainable fishing practices, and screens for heavy metals.
Vitamin B complex	1 capsule daily	Dr. Brighten B-Active Plus or Integrative Therapeutics Active-B Complex	Choose a supplement with activated B vitamins such as methylfolate, B12 (methylcobalamin), and B6 (riboflavin 5'-phosphate). Avoid folic acid.

14-DAY LIVER DETOX

Supplement Type	How to Take	Recommended Brands	Notes
Liver support powder	1 drink mix packet twice daily Take for the first 14 days of the program. Repeat every 3–4 months if staying on the pill.	Included in Dr. Brighten Paleo Detox or Dr. Brighten Plant-Based Detox kits, or Designs for Health VegeCleanse	Pea protein–based detox powder requires additional digestive enzyme support. This is included in the Plant-Based Detox kit or you can use Designs for Health Hydrolyzyme. Dr. Brighten kits contain all essential supplements.

Brighten Supplement Protocol

14-DAY LIVER DETOX

Supplement Type	How to Take	Recommended Brands	Notes
Phase II liver detox support (glutamine, glycine, taurine, alpha-ketoglutarate, glutathione, methione, ornithine)	3 capsules twice daily during 14-day detox. Can continue for additional liver and gallbladder support.	Included in Dr. Brighten Paleo Detox or Plant-Based Detox kits, Designs for Health Amino-D-Tox	

GUT REPAIR

Supplement Type	How to Take	Recommended Brands	Notes
Gut repair herbs and nutrients (L-glutamine, DGL, aloe vera, slippery elm, chamomile, marshmallow root, cat's claw, quercetin, zinc carnosine, N-acetyl-D-glucosamine, citrus pectin, MSM)	3 capsules with breakfast 4 capsules with dinner Take for 30–90 days.	Dr. Brighten Gut Rebuild or Xymogen GlutAloeMine	Avoid shellfish-derived N-acetyl-D-glucosamine if you have a shellfish allergy. The Dr. Brighten brand does not contain shellfish.
Comprehensive digestive support	2 capsules daily with meals	Dr. Brighten Digest or Xymogen XymoZyme	Look for a supplement with digestive enzymes, hydrochloric acid, and bile acid for complete digestive support.
Grass-fed collagen	2 tablespoons twice daily	Great Lakes Gelatin, Bulletproof Collagen Protein, or Vital Proteins Collagen Peptides	Choose pasture-raised, grass-fed, and organic.

Brighten Supplement Protocol

GUT REPAIR

Supplement Type	How to Take	Recommended Brands	Notes
Turmeric	500–1,000 mg daily	Dr. Brighten Turmeric Boost or Integrative Therapeutics Curcumax Pro	Look for high bioavailable sources of turmeric with active ingredients for best results.
Lactobacillus and *Bifidobacterium* probiotic	50 billion CFUs daily After the 30-day program, 15–20 billion CFUs daily	Klaire Labs Ther-Biotic Complete or Ther-Biotic Factor 4	If you suspect that you have SIBO, skip the *Lactobacillus* and instead begin with *Saccharomyces boulardii* and a spore-forming probiotic.
Saccharomyces boulardii probiotic	500–2,000 milligrams daily	Designs for Health FloraMyces	Beneficial yeast

CATEGORY A: TOO MUCH ESTROGEN

Supplement Type	How to Take	Recommended Brands	Notes
Turmeric	500–1,000 mg daily	Dr. Brighten Turmeric Boost or Integrative Therapeutics Curcumax Pro	Look for high bioavailable sources of turmeric with active ingredients for best results.
Melatonin	0.5–3 mg nightly	Dr. Brighten Sweet Dreams or Designs for Health Melatonin	If you wake feeling groggy, take one hour before bed.

Brighten Supplement Protocol

CATEGORY A: TOO MUCH ESTROGEN

Supplement Type	How to Take	Recommended Brands	Notes
Comprehensive hormone support (B6, B12, folate, DIM, broccoli extract, calcium-D-glucarate, green tea extract, black cohosh, Vitex, resveratrol, magnesium, chrysin)	2 capsules twice daily	Dr. Brighten Balance or Integrative Therapeutics Femtone and Indolplex with DIM	Choose a supplement that supports estrogen metabolism in the liver and the gut.

CATEGORY B: TOO LITTLE ESTROGEN

Supplement Type	How to Take	Recommended Brands	Notes
Maca, gelatinized powder	5 g daily, perhaps added to a smoothie	Gaia Herbs, Femmenessence, or Dr. Jess Hormone Master	Look for gelatinized maca in powder form.
Black cohosh	100 mg daily or ½–1 teaspoon tincture once or twice daily Use days 1 through 14 of your cycle, or new moon to full moon.	Dr. Brighten Balance (2 caps twice daily) or Wise Woman Herbals (tincture)	Estrogen-supportive herb
Vitamin E	400 IU daily Take for at least 30 days.	Integrative Therapeutics Vitamin E	Look for 400 IU of d-alpha tocopherol.

Brighten Supplement Protocol

CATEGORY B: TOO LITTLE ESTROGEN

Supplement Type	How to Take	Recommended Brands	Notes
Magnesium bisglycinate	300–600 mg daily	Dr. Brighten Magnesium Plus or Klaire Labs Magnesium Glycinate Complex	Citrate is an alternative form of magnesium that can help with constipation.

CATEGORY C: TOO LITTLE PROGESTERONE

Supplement Type	How to Take	Recommended Brands	Notes
Vitex (chaste tree berry)	200 mg daily or 1–2 teaspoons twice daily day 15 through day 28 of your cycle, or full moon to new moon.	Dr. Brighten Balance or Wise Woman Herbals	Most of my patients prefer the capsule form because of the strong taste Vitex tincture has.
Vitamin C	5 mL daily liposomal or 1,000–4,000 mg buffered vitamin C	Designs for Health Liposomal Vitamin C or Integrative Therapeutics Buffered Vitamin C	Liposomal is a highly absorbable form. Vitamin C can cause loose stools at higher doses.
Vitamin B complex	1 capsule daily	Dr. Brighten B-Active Plus or Integrative Therapeutics Active B-Complex	Choose a supplement with activated B vitamins such as methylfolate, B12 (methylcobalamin), and B6 (riboflavin 5'-phosphate). Avoid folic acid.

Brighten Supplement Protocol

CATEGORY C: TOO LITTLE PROGESTERONE

Supplement Type	How to Take	Recommended Brands	Notes
Bioidentical progesterone	As directed by your doctor	Bioidentical created at a compounding pharmacy	Bioidentical progesterone is not the same as progestin. I recommend avoiding progestin due to high side effects.

CATEGORY D: TOO MUCH TESTOSTERONE

Supplement Type	How to Take	Recommended Brands	Notes
Omega-3	2 capsules once or twice daily	Dr. Brighten Omega Plus or Nordic Naturals ProEPA Xtra	Look for brands that use third-party testing, sustainable fishing practices, and screens for heavy metals.
Comprehensive testosterone support (B6, vitamin D, zinc, saw palmetto, nettle, L-glycine, L-alanine, chrysin, DIM, lycopene)	2 capsules daily	Dr. Brighten Saw Palmetto Plus or Integrative Therapeutics Pros-Forte	I recommend testing to screen for DHT production and monitoring symptoms.
Comprehensive hormone support (B6, B12, folate, DIM, broccoli extract, calcium-D-glucarate, green tea extract, black cohosh, Vitex, resveratrol, magnesium, chrysin)	2 capsules twice daily	Dr. Brighten Balance or Integrative Therapeutics Femtone and Indolplex with DIM	Choose a supplement that supports hormones and metabolism in the liver and gut.

Brighten Supplement Protocol

CATEGORY D: TOO MUCH TESTOSTERONE

Supplement Type	How to Take	Recommended Brands	Notes
Comprehensive adrenal support (vitamin C, B2, B6, pantothenic acid, *Eleuthero*, American ginseng, ashwagandha, Rhodiola, N-acetyl tyrosine, licorice root)	3 capsules in the morning	Dr. Brighten Adrenal Support or Integrative Therapeutics HPA Adapt plus Active B-Complex and Vitamin C	Avoid licorice if you have a history of high blood pressure.
Inositol	2 g once or twice daily	Designs for Health	Supports women's hormones and promotes restful sleep.

CATEGORY E: TOO LITTLE TESTOSTERONE

Supplement Type	How to Take	Recommended Brands	Notes
Comprehensive adrenal support (vitamin C, B2, B6, pantothenic acid, *Eleuthero*, American ginseng, ashwagandha, Rhodiola, N-acetyl tyrosine, licorice root)	3 capsules in the morning	Dr. Brighten Adrenal Support or Integrative Therapeutics HPA Adapt plus Active B-Complex and Vitamin C	Avoid licorice if you have a history of high blood pressure.
DHEA (topical)	10–15 mg daily	Compounded by your doctor or Julva	Avoid contact with people and pets when cream is applied. Stop taking DHEA if you develop oily skin.

Brighten Supplement Protocol

CATEGORY F: TOO LITTLE CORTISOL

Supplement Type	How to Take	Recommended Brands	Notes
Comprehensive adrenal support (vitamin C, B2, B6, pantothenic acid, *Eleuthero*, American ginseng, ashwagandha, Rhodiola, N-acetyl tyrosine, licorice root)	3 capsules in the morning	Dr. Brighten Adrenal Support or Integrative Therapeutics HPA Adapt plus Active B-Complex and Vitamin C	Avoid licorice if you have a history of high blood pressure.
Vitamin C	5 mL daily liposomal or 1,000–4,000 mg buffered vitamin C	Designs for Health Liposomal Vitamin C or Integrative Therapeutics Buffered Vitamin C	Liposomal is a highly absorbable form. Vitamin C can cause loose stools at higher doses.
Vitamin B complex	1 capsule daily	Dr. Brighten B-Active Plus or Integrative Therapeutics Active B-Complex	Choose a supplement with activated B vitamins such as methylfolate, B12 (methylcobalamin), or B6 (riboflavin 5'-phosphate). Avoid folic acid.

Brighten Supplement Protocol

CATEGORY G: TOO MUCH CORTISOL

Supplement Type	How to Take	Recommended Brands	Notes
Comprehensive adrenal support (vitamin C, B2, B6, pantothenic acid, *Eleuthero*, American ginseng, ashwagandha, Rhodiola, N-acetyl tyrosine, licorice root)	3 capsules in the morning	Dr. Brighten Adrenal Support or Integrative Therapeutics HPA Adapt plus Active B-Complex and Vitamin C	Avoid licorice if you have a history of high blood pressure.
Comprehensive cortisol-lowering formula (vitamins C, B1, B2, B6, B12, pantothenic acid, magnesium, taurine, L-theanine, lemon balm, passionflower, Valerian, ashwagandha, phosphatidylserine)	3 capsules at bedtime	Dr. Brighten Adrenal Calm or Wise Woman Herbals Valerian Compound plus Designs for Health Phosphatidylserine Powder, Vitamin C, and B-Supreme	The herbs listed promote a calm nervous system.
Vitamin B complex	1 capsule daily	Dr. Brighten B-Active Plus or Integrative Therapeutics Active B-Complex	Choose a supplement with activated B vitamins such as methylfolate, B12 (methylcobalamin), and B6 (riboflavin 5'-phosphate). Avoid folic acid.

Brighten Supplement Protocol

CATEGORY G: TOO MUCH CORTISOL

Supplement Type	How to Take	Recommended Brands	Notes
Vitamin C	5 mL daily liposomal or 1,000–4,000 mg buffered vitamin C	Designs for Health Liposomal Vitamin C or Integrative Therapeutics Buffered Vitamin C	Liposomal is a highly absorbable form. Vitamin C can cause loose stools at higher doses.

CATEGORY H: TOO LITTLE THYROID HORMONE

Supplement Type	How to Take	Recommended Brands	Notes
Comprehensive thyroid support formula (vitamin A, B2, iodine, zinc, selenium, copper, manganese, chromium, N-acetyl tyrosine, American ginseng, forskolin extract)	2 capsules daily	Dr. Brighten Thyroid Support	Never take iodine as a stand-alone supplement without selenium or consulting your doctor.
Comprehensive adrenal support (vitamin C, B2, B6, pantothenic acid, *Eleuthero*, American ginseng, ashwagandha, Rhodiola, N-acetyl tyrosine, licorice root)	3 capsules in the morning	Dr. Brighten Adrenal Support or Integrative Therapeutics HPA Adapt plus Active B-Complex and Vitamin C	Avoid licorice if you have a history of high blood pressure.
Omega-3	2 capsules once or twice daily	Dr. Brighten Omega Plus or Nordic Naturals ProEPA Xtra	Look for brands that use third-party testing, sustainable fishing practices, and screens for heavy metals.

Your Custom Supplement Table

Fill in this table with the specific supplements you will be taking to fit your individualized needs:

	Supplement Type	Dose	Brand
Morning			
Afternoon			
Evening			

Liver Detox

For the liver detox, you will follow either a bone broth–based detox or a pea protein detox if you're a vegetarian. (Please note that if you have SIBO, then don't follow the pea protein detox, as it will likely cause bloating.) Start the liver detox on Day 1 of the 30-day program and continue through Day 14 (see Brighten Supplement Protocol). You will consume one of the powder packets twice a day, which is dense with nutrients and helps your liver replenish what has been lost. You can add these packets to your morning smoothie or simply to water or a nondairy milk. You will also take antioxidants and amino acids in supplement form. Antioxidants are anti-inflammatory and help with mood, metabolic issues, and painful periods; amino acids are also essential for proper liver function. Many of my patients take these supplements in the midafternoon if they start to feel tired, because the B vitamins give them that boost to keep going. Now, let's rock this detox!

30-Day Brighten Protocol Diet

For the next 30 days, eliminate all gluten and grains, dairy, soy, sugar, caffeine, alcohol, and inflammatory fats. Follow the meal plans and recipes each day to ensure that you're nourishing your body with the right vegetables, protein, and healthy fats. I have specifically designed these meal plans to incorporate gut-promoting foods, as well as liver-loving options during the first half of the program. Don't forget to drink plenty of water—especially during the first two weeks, when you'll be detoxing. Drink half your body weight in fluid ounces daily plus an additional 20 ounces during the detox.

Seed Cycling

Yes, seeds can be used to influence your hormones and correct or eliminate irregular or heavy periods, PMS, infertility, and symptoms of perimenopause and menopause. You will focus on certain seeds in the first half of your cycle to increase estrogen and other seeds in the second half to increase progesterone, supporting your natural hormone rhythms. To seed cycle, add the following to your diet (follow the moon cycle if you don't have a period):

- Day 1 to Day 14 (new moon to full moon): During the follicular phase, you need more estrogen in order to build up your uterine lining. Eat 2 tablespoons of freshly ground flaxseeds and pumpkin seeds daily to increase your estrogen level.
- Day 15 to Day 30 (full moon to new moon): During the luteal phase, the corpus luteum (what is left behind after the egg ruptures) begins to release progesterone, which helps thicken the uterine lining, readying the body for a baby. (No worries if you don't want a baby; seeds are still your friend.) Eat 2 tablespoons of freshly ground sesame seeds and sunflower seeds daily to boost your progesterone production.

You can use seed cycling even if you're not menstruating or you're on the pill by following the moon cycle. (See page 230 for more moon info.)

The 30-Day Plan

I've included a sample day of the program for you to follow on Day 1 (with the lifestyle practices, supplements, and so on). You will repeat this routine during all the other days of the program except your meals will change. I've provided you with meal plans that include quick and easy-to-make recipes, which taste delicious too. These meals are designed to optimize your hormones and your detox system, and increase your energy. Please note that since many of the meals make more than one serving, you will eat the remaining servings on later days. Tailor this program to fit your particular hormone imbalance, and be sure to include the liver detox packets and supplements from Day 1 through Day 14.

Sample Day 1

MORNING

- When you wake up, take your temperature if you're using a fertility awareness method (FAM). Take five deep belly breaths, then expose yourself to natural light, either by opening up the curtains or going outside for at least five minutes. If you can get your bare feet on the ground and do some "earthing," even better. You can also purchase a natural-light alarm clock, which will help expose you to light that's similar to sunshine.

- Drink 8 to 16 ounces of warm lemon water.

- Read your journal. Take five minutes to engage in some kind of relaxation practice, such as meditating, prayer, breath work, yoga, or HeartMath to set your parasympathetic tone for the day. Write in your journal about any experiences you may have had during your relaxation practice.

- Take ten minutes to move. Find a yoga app or Pilates video on YouTube—or whatever is your preference—and engage your body in some morning movement.

- Look at your calendar to see what day it is on the moon/menstrual cycle.

- Be prepared to take note of the signs and symptoms your body is providing you.

BREAKFAST

- Start your day with a Morning Matcha Smoothie (page 293). (You can make this the night before for a grab-and-go option too.)

- Take your supplements: a prenatal or multivitamin and B complex, the probiotics, and the omega-3, plus any of the supplements that apply to your hormone imbalance. On this first day, also take the morning packet of Paleo or Plant-Based Detox powder.

Stress-Reducing Practices

The following tools can help you master your stress response and rebalance your hormones during the program. Self-care is not selfish. It is essential to your health.

☐ **Meditate.** Take five minutes every morning to tune in to your breath, your body, and your thoughts. Release the thoughts that no longer serve you, and set your intentions for the day. Incorporate deep breathing to engage the "rest and digest" aspect of the nervous system. Try a meditation class or grab yourself a meditation device like Muse to help you master your practice.

☐ **Breathe deeply.** Aim for fifty to one hundred cleansing breaths throughout the day, with an emphasis on creating a longer exhale to help nourish the parasympathetic ("rest and digest") nervous system. Set the alarm on your smartphone to remind you to do this every hour and at the start and end of every day. Deep breathing will relax your mind and help your lungs eliminate metabolic waste.

☐ **Keep a gratitude journal.** Each night write down the three things you're grateful for and then set your intention for the next day, writing it down too. When you wake in the morning, take five deep belly breaths, then read your journal before anything else, looking at the intention you set for yourself the night before. This ensures you own your day and that you decide the outcome.

☐ **Track your progress.** Congratulate yourself on your accomplishments during the program, and notice where you veer off track. Set aside a few minutes at the end of each day to record any difficulties you've encountered, along with the physical and emotional changes you've seen. How do you feel? How is your mood? Are you less irritable or more energized? Does your skin look better? Do you feel less bloated? It's important for you to celebrate the positive effects these changes have on your body.

☐ **Visualize.** You can make a vision board or simply close your eyes, think about your goals, and then envision what it feels like to reach those goals. This practice will not only keep you motivated but also decrease your stress and help you get organized, both of which your hormones love!

☐ **Pray.** Grow your spiritual practice with prayer if that feels right for you.

☐ **Practice mindfulness.** Focus on being present and observing your surroundings without judgment for just five minutes a day. This can help decrease anxiety, stress, and depression.

☐ **Find your happy place.** Visualize a time or place that brings you happiness. Engage all your senses. Remember the smells that surrounded you, the foods you ate, the feel of the warm breeze. Use your imagination to take a virtual trip when you feel super stressed out. If you can't take that beach vacation you've been dreaming of right now, at least you can pretend you're there and experience some of the relaxation benefits.

☐ **Rest in a float tank.** Submerging your body in a float tank can be deeply relaxing.

☐ **Indulge in a massage, acupuncture, or Reiki.** All of these practices can be great ways to pamper yourself and reduce stress.

☐ **Get a mani-pedi.** Taking time for self-care is really important when trying to reduce stress, not to mention the relaxing massages that usually accompany a mani-pedi.

☐ **Dance.** Go out dancing with your tribe. This is a great way to blow off some steam and get some movement at the same time.

☐ **Play.** Make time to play, either by yourself, with your pet, or with a friend. Do something *fun* every day. Nurture your inner child, who wants to have some fun. Laughter is essential for stress management and a healthy heart.

☐ **Practice art.** Paint, draw, sculpt, sew, knit, or engage in whatever creative activity you enjoy. When we zone out during creative activities it relieves stress, because our minds are focused on the activity instead of our to-do list.

☐ **Have sex.** You knew this was coming, right? Orgasms are good for your health and a great form of stress release, so try to have at least one per week during the program.

MIDMORNING

- Enjoy a warm beverage, like the Restorative Roots Liver Tonic (page 306).

LUNCH

- Enjoy the Curly Kale with Cilantro Artichoke Pesto for lunch (page 309). Take any lunchtime supplements too.

- After lunch, take a ten-minute walk or do some other form of exercise.

AFTERNOON

- Remind yourself which seed you should be eating as part of your seed cycling, and as an afternoon snack, enjoy either one of the Flax and Pumpkin Balls (page 301) or one of the No-Bake Sunflower Rose Cookies (page 302).

- If you need a pick-me-up in the afternoon, this is a good time to enjoy an herbal tea or mushroom elixir (check out "Caffeine-Free Coffee Replacements," page 122) or drink more water.

- Depending upon your physical activity for the day so far, do ten to twenty minutes of Pilates, yoga, swimming, walking, playing with your kids, weight lifting, high-intensity interval training, hula-hooping, or any other exercise you enjoy.

DINNER

- Enjoy Whole Chicken with Aromatics (page 328) with steamed broccoli for dinner.

- Take your multivitamin or prenatal, probiotics, omega-3, and any other evening supplements, plus your Paleo or Plant-Based Detox powder.

- If you like, make this your one night of the week to go by candlelight, and light those candles, girl!

EVENING

- At least two hours before bedtime, put on your amber glasses—and if you're going to be working on a computer, make sure you have the option to use the blue-light blocking Night Shift or f.lux.

Upgraded Golden Milk

I like to use Upgraded Golden Milk (recipe on page 304) at night because it's a combination of turmeric, fat, and collagen, and that collagen is rich in glycine, which is going to help you feel nice and relaxed as you go to bed at night. The anti-inflammatory turmeric supports healthy cortisol levels and the fat will help your body feel satiated, so you won't get that blood sugar spike. If you're someone who often wakes up between 2 and 4 a.m., then drinking this beverage two hours before bed can actually change your sleep within three to five days.

- Before bed, relax with some meditation, deep breathing, visualizations, or journaling.
- Drink a relaxing caffeine-free herbal tea or an upgraded golden milk (see the above box).

BEDTIME

- Remember to darken your bedroom for sleep and keep the temperature below 70 degrees Fahrenheit, plus turn off all electronics.
- Sleep at least eight hours. If typically your children or pets wake you during the night, try using either a red headlamp or a pink Himalayan salt lamp to avoid disrupting your melatonin.

Day 2

    Breakfast: Piña Colada Cleanser (page 293)
    Lunch: Curly Kale with Cilantro Artichoke Pesto (leftover) with Chicken with Aromatics (leftover)
    Dinner: Sesame Carrot and Cabbage Buffalo Stir-Fry (page 325)

Day 3

    Breakfast: Morning Matcha Smoothie (page 293)
    Lunch: Mango Chicken Collard Wraps with Golden Curry Sauce (page 319)
    Dinner: Red Curry Salmon (page 321) served over cauliflower rice or with a green salad

Day 4

Breakfast: Piña Colada Cleanser (page 293)
Lunch: Red Curry Salmon (leftover) with a spinach salad
Dinner: Mango Chicken Collard Wraps with Golden Curry
Sauce (leftover)

Day 5

Breakfast: Upbeet Citrus Smoothie (page 295)
Lunch: Tri-Color Cabbage Slaw (page 312)
Dinner: Sesame Carrot and Cabbage Buffalo Stir-Fry (leftover)

Day 6

Breakfast: Creamy Strawberry Cauliflower Smoothie (page 291)
Lunch: Tri-Color Cabbage Slaw (leftover)
Dinner: Garlic Shrimp over Chili Lime Cauliflower Rice
(page 316)

Day 7

Breakfast: Upbeet Citrus Smoothie (page 295)
Lunch: Baby Bok Choy Salad with Chickpea Miso (page 307) and
Garlic Shrimp (leftover)
Dinner: Seared Fish with Tomatoes and Capers (page 324)

Day 8

Breakfast: Creamy Strawberry Cauliflower Smoothie (page 291)
Lunch: Baby Bok Choy Salad with Chickpea Miso and Seared
Fish (leftover)
Dinner: Cauliflower Tabbouleh (page 308) with protein of choice
or fresh avocado slices

Day 9

Breakfast: Spicy Carrot Cleanser Smoothie (page 294)
Lunch: Cauliflower Tabbouleh (leftover)
Dinner: Sardine Fritters (page 323) with a green salad

Day 10

Breakfast: Lemon Berry Boost (page 292)
Lunch: Liver-Cleansing Beet Slaw (page 310)
Dinner: Whole Chicken with Aromatics (page 328) with
cauliflower rice

Day 11

Breakfast: Spicy Carrot Cleanser Smoothie (page 294)
Lunch: Sardine Fritters (leftover) with Liver-Cleansing Beet
Slaw (leftover)
Dinner: Lemongrass Thai Chicken Soup (page 318; use leftover
Chicken with Aromatics)

Day 12

Breakfast: Lemon Berry Boost (page 292)
Lunch: Chicken with Aromatics (leftover) with steamed broccoli
Dinner: Mediterranean Lamb Sliders (page 321)

Day 13

Breakfast: Morning Matcha Smoothie (page 293)
Lunch: Mediterranean Lamb Sliders (leftover)
Dinner: Lemongrass Thai Chicken Soup (leftover)

Day 14

Breakfast: Piña Colada Cleanser (page 293)
Lunch: Thai Zoodles with Citrus Almond Sauce (page 311)
Dinner: Dijon and Almond Herb-Crusted Salmon (page 315) with
cauliflower rice or a green salad

Day 15

Breakfast: Creamy Strawberry Cauliflower Smoothie (page 291)
Lunch: Dijon and Almond Herb-Crusted Salmon (leftover) with a
green salad
Dinner: Thai Zoodles with Citrus Almond Sauce (leftover)

Day 16
Breakfast: Shiitake Tarragon Mini Frittatas (page 297)
Lunch: Curly Kale with Cilantro Artichoke Pesto (page 309)
Dinner: Zucchini Turkey Burgers (page 329) with greens

Day 17
Breakfast: Morning Matcha Smoothie (page 293)
Lunch: Zucchini Turkey Burgers (leftover) with greens
Dinner: Curly Kale with Cilantro Artichoke Pesto (leftover)

Day 18
Breakfast: Shiitake Tarragon Mini Frittatas (leftover)
Lunch: Tri-Color Cabbage Slaw (page 312) with protein of choice
Dinner: Garlic Shrimp over Chili Lime Cauliflower Rice
    (page 316)

Day 19
Breakfast: Piña Colada Cleanser (page 293)
Lunch: Garlic Shrimp over Chili Lime Cauliflower Rice (leftover)
Dinner: Red Curry Salmon (page 321) with steamed broccoli

Day 20
Breakfast: Brussels Sprouts Breakfast Hash (page 296)
Lunch: Tri-Color Cabbage Slaw (leftover)
Dinner: Ginger-Marinated Cod (page 317) with steamed
    bok choy

Day 21
Breakfast: Upbeet Citrus Smoothie (page 295)
Lunch: Cauliflower Tabbouleh (page 308)
Dinner: Red Curry Salmon (leftover) with a green salad

Day 22
Breakfast: Brussels Sprouts Breakfast Hash (leftover)
Lunch: Cauliflower Tabbouleh (leftover)
Dinner: Ginger-Marinated Cod (leftover) with sautéed Swiss chard

Day 23

    Breakfast: Creamy Strawberry Cauliflower Smoothie (page 291)
    Lunch: Liver-Cleansing Beet Slaw (page 310)
    Dinner: Citrus-Marinated Flank Steak (page 314) with a green salad

Day 24

    Breakfast: Sweet Carrot Breakfast Patties (page 299)
    Lunch: Citrus-Marinated Flank Steak (leftover) with Liver-Cleansing Beet Slaw (leftover)
    Dinner: Whole Chicken with Aromatics (page 328) with steamed broccoli

Day 25

    Breakfast: Spicy Carrot Cleanser Smoothie (page 294)
    Lunch: Mango Chicken Collard Wraps with Golden Curry Sauce (page 319)
    Dinner: Dijon and Almond Herb-Crusted Salmon (page 315) with a green salad

Day 26

    Breakfast: Spinach Sage Breakfast Patties (page 298)
    Lunch: Dijon and Almond Herb-Crusted Salmon (leftover) with a green salad
    Dinner: Tikka Masala Turkey Meatballs (page 326) with cauliflower rice

Day 27

    Breakfast: Sweet Carrot Breakfast Patties (leftover)
    Lunch: Mango Chicken Collard Wraps with Golden Curry Sauce (leftover)
    Dinner: Mediterranean Lamb Sliders (page 321)

Day 28

    Breakfast: Lemon Berry Boost (page 292)
    Lunch: Mediterranean Lamb Sliders (leftover)
    Dinner: Tikka Masala Turkey Meatballs (leftover)

Day 29
Breakfast: Spinach Sage Breakfast Patties (leftover)
Lunch: Sardine Fritters (page 323) with a green salad
Dinner: Thai Zoodles with Citrus Almond Sauce (page 311)

Day 30
Breakfast: Morning Matcha Smoothie (page 293)
Lunch: Thai Zoodles with Citrus Almond Sauce (leftover)
Dinner: Sardine Fritters (leftover) with cauliflower rice

The Transition Phase: Food Reintroductions

Congratulations on completing your 30-day program! You go, girl! I'm so proud of you and excited for you to transition into a hormone-loving lifestyle. After you complete the 30 days of this program, the next step will be to reevaluate your hormone experience and your hormone symptoms. You will take the Hormone Quiz again to discover what improved during the 30 days and what areas are still in need of focus. Compare it to the first quiz you took before you started the program, and see if you notice any changes, then reevaluate your symptoms. Did you check fewer boxes in a certain category after 30 days? If you don't see a huge shift in your hormones, *do not* feel discouraged. While most people experience improvement in symptoms, some women need to have a full menstrual cycle before they notice a significant change.

Remember, it can take a couple of months to really shift your hormones and see noticeable changes. Keep going with this style of eating to continue your progress. Post–birth control syndrome lasts for an average of at least four to six months in most women, so it makes sense that your body will need support during that time.

After the 30 days, you'll begin a food reintroduction phase, inviting back in some of those foods you eliminated during the program, so you can discover how these foods affect you. The reintroduction is absolutely everyone's favorite part of the program; bringing back in the foods you love and have been craving can feel super satisfying. Reintroducing the foods you eliminated can add more variety to your diet, but only if they work for you.

As you move forward with food reintroductions, I don't want you to abandon any of the great dietary habits you will have picked up in the program. For example, healthy fats (like avocado, avocado oil, olive oil, and macadamia nut oil) are the backbone of your hormones—you can't make hormones without fat. And continue to eat lots of vegetables, since they will support your liver as you move forward with food reintroduction. As for grains and gluten, dairy, soy, sugar, caffeine, alcohol, and inflammatory fats, because they are the most reactive foods for the majority of women with hormone imbalances, you will bring them back in one at a time, a few days apart, to test and understand if these foods work for you.

Start your reintroduction phase first with caffeine, then dairy, then soy, then grains, one at a time. Reintroduce coffee first because (1) you're probably jonesing for some java, and (2) you're going to know pretty quickly whether or not it will work for you. Reintroducing proteins like dairy and soy, which are top allergens, can definitely aggravate you—and those symptoms may not go away for several weeks. If you have acne, give yourself three to six months before reintroducing dairy. And if you're going to reintroduce soy, it better be organic. You don't want to mess around with any GMOs. Don't reintroduce inflammatory fats—they are never good for your body and there's no point in bringing them back into your diet. You're in it to win it, lady!

What about sugar and alcohol? Do you even want to bring these back in? Okay, maybe you *want* to, but here's why you should reconsider. One drink alone can cause an imbalance in estrogen and progesterone. Because of the pressure that both sugar and alcohol place on your entire system, these foods can contribute to hormonal imbalance and increase inflammation, not to mention mess with your blood sugar. Any time you have a blood sugar imbalance, you run into trouble with your hormones. The fluctuations create hormonal chaos in the body and keep you from feeling well. So reapproach sugar with caution. If you're going to bring it back in, stay away from high-fructose corn syrup or other highly processed sugars. If you're going to reintroduce alcohol, I recommend having one alcoholic drink and observing your symptoms throughout the entire week. Frequently I will have women track an entire menstrual cycle without alcohol and then reintroduce it in their next cycle and track their symptoms. Often periods get more difficult when your liver is burdened with things like alcohol. So be cautious and perhaps treat sugar and alcohol as indulgences, not everyday items. Also, I don't recommend reintroducing alcohol if you have estrogen dominance. Does this mean you can never drink? Of course not! But it does mean that when you attend a party, event, or other social function and feel inclined to have a drink, your body will be more likely to recover and recalibrate more quickly.

For a successful reintroduction process, as I said, take them one at a time, track, observe, and rotate. For example, when you reintroduce dairy, have dairy products two to four times in a single day and then stop eating dairy for the following two days. Wait and watch for symptoms, and be patient. Track your symptoms using the "Food Reintroduction Symptom Tracker" (page 271). What should you look for? The return of any previous concerns as well as possibly new symptoms. Any symptoms that disappeared during the 30-day plan should not be coming back. If they are, you need to remove that food from your diet ASAP, and I recommend removing it for at least three months to give your body time to recalibrate. You may then reintroduce it again to see if your body will handle it differently. The one exception I see in my practice is gluten. If you reintroduce gluten and have symptoms, it may be a food you have to say goodbye to for good.

What are the symptoms of a food sensitivity? You can experience a range of reactions, anything from headaches to digestive issues like constipation, heartburn, bloating, or diarrhea to skin symptoms like acne or rashes. Feeling bloated or fatigued after eating, seeing shifts in your mood, or sleep disturbances can also be signs of a food sensitivity. The worse you feel after eating that food and the longer your symptoms last, the longer you should wait before testing that food again. Move forward with a reintroduction only when your symptoms have cleared and you're feeling relatively back to normal.

Common Symptoms of Food Sensitivity

acne	bloating	body aches
coughing	dizziness	fatigue
headaches	heartburn	indigestion
mood changes	rashes	sinus congestion
sleep issues	water retention	

Food Reintroduction Symptom Tracker

This is an opportunity to be your own health detective when it comes to your diet. Remember, reintroduce only one new food at a time. On day 1 of the reintroduction, eat that food two to four times in the same day. Stop eating it, then wait three days to see if you have a reaction. Track your response each day, noting your symptoms below. If you have no reaction to the food, you can keep that food in your food plan, and continue with the next food for reintroduction. If you are unsure whether you had a reaction, retest the same food in the same manner. If you have a reaction, remove that food from your diet and reintroduce it at a later date.

	Day 1	Day 2	Day 3	Day 4
Food				
Digestion or bowel changes				
Joint or muscle aches				
Headache or pressure				
Nasal or chest congestion				
Changes in urination				
Skin				
Energy level				
Sleep				
PMS				
Mood				
Cramps				
Other symptoms				

After you've successfully reintroduced a food, eat that food in a rotating fashion. Aim for eating it no more than four days a week. This rotation is so you don't develop a food sensitivity by consuming too much of it. Essentially, you want to make it a habit to not consume the same foods day in and day out. This is about finding what works for *your* body, discovering the foods that serve you best and the foods that might not work for you right now.

Life After the 30-Day Brighten Program

Has your dress size changed or are your skinny jeans fitting better? Are you finding that your period was easier, you had fewer cramps, or you no longer get hormonal headaches? Are you experiencing less fatigue, more energy, and sleeping soundly through the night? How does your skin look? How does your hair look? Are you feeling like your digestion is a lot better and you have fewer cravings? Have you gotten rid of your constipation, bloating, gas, or diarrhea? Has your mood improved—are you less depressed, anxious, or irritable? Are you handling stress much better and using mind–body tools for mastering your stress response? Has your libido come out of hiding? What has resolved for you? Now that you've finished the 30-day program, reflect on how you feel and congratulate yourself. Think about what has improved and what still needs work, as well as which lifestyle habits you want to continue.

If you feel you need support while doing a hormone-reboot program like the 30-Day Brighten Program, either during the program or after the 30 days, honor that. You can work with me or my clinicians in our clinic, or visit our online community and take advantage of the resources there to support you in your next phases (visit DrBrighten.com/Resources).

I hope you now feel more empowered to take charge of your health, investigate the root cause of your symptoms, and make the best decisions for you. Here's to happy hormones!

CHAPTER 13

ALTERNATIVE BIRTH CONTROL METHODS

I want to begin this chapter by acknowledging that the only person who truly understands your body and what it needs is *you*. You are the only one who can determine whether or not the benefits of taking the pill outweigh the risks. If I've done my job, this book has given you the information you need to make that decision.

I believe doctors should teach women how their hormones work and help them understand which signs and symptoms are important. One method I encourage all women to learn is the fertility awareness method (FAM). Now, many people—including doctors—are often quick to dismiss FAM, claiming women are too lazy or too easily confused by the workings of their own bodies to successfully use this method. I disagree. In fact, statistics show that *one in five* women wishes to learn more about FAM, which is why I present it here in this chapter and explain how to use it. In my opinion, women have tremendous motivation to understand their bodies and prevent pregnancy. But beyond preventing pregnancy, FAM is a valuable tool for all women working toward mastering their hormones. Unfortunately, many women don't fully learn the details of their cycles until they're trying to get pregnant and want to pinpoint ovulation.

Let's just take a moment to recognize that while your doc may have told you the pill was your best bet in preventing pregnancy, it isn't 100 percent effective. If you look at typical use of the birth control pill, or how you usually take it, which is not the same exact time every day—sometimes you miss it, sometimes you might have an upset stomach, sometimes you might have diarrhea, and sometimes you might accidentally take it with alcohol (well, maybe not accidentally)—it's about 91 percent effective, meaning 9 out of every 100 women who take it get pregnant. All these variables affect your absorption and how well that birth control pill actually works. Not to mention there are interactions with other medications. And despite all the hormonal birth control options we have, almost half of all pregnancies in the US in 2011 were unintended, according to a study published in the *New England Journal of Medicine*.

Are you ready to ditch the pill? If the answer to that question is yes, this chapter will provide you with a comprehensive look at alternative methods of pregnancy prevention. I am not suggesting that any of these methods will work for every woman, but these non-hormonal alternatives are worth exploring to understand the best method for you.

The Fertility Awareness Method

FAM is a non-hormonal approach to birth control designed to teach you how to recognize the signs of ovulation based on your body's signals. It's based on the following concepts:

- You ovulate only one egg in each menstrual cycle.

- The egg lives for no more than twenty-four hours.

- Sperm are patient little creatures that hang around for several days looking to impregnate you. They can live five to six days in the uterus.

- Add all of this up and you're potentially fertile for a max of eight days out of your cycle (but typically six).

If you're not looking to get into the baby-making business any time soon, then you will need to either use another means of contraception or abstain from sex during your eight-day ovulatory window, which begins about six days before ovulation, includes the day of ovulation, and lasts through the day after ovula-

In This Chapter

- How to use the fertility awareness method (FAM)

- Just how effective non-hormonal birth control options are

- Clearing up condom confusion

- How diaphragms and sponges work

- Is the pull-out method legit?

tion. Some methods increase that window to more than eight days in order to reduce the risk of pregnancy. To ensure there is no chance of pregnancy, most people elect to use a condom or another form of contraception at this time.

Even if you don't think FAM is right for you, consider at least using it as a means of getting in touch with your body and its internal rhythms. This can enable you to identify where you may need to make shifts to create hormonal harmony in your body and your life.

Does FAM Really Work?

You may be skeptical about whether or not the fertility awareness method actually works. If you're used to relying on a pill that stops you from ovulating, trusting your ability to master your body's fertile signs and avoiding sex during the times when you're the most fertile can seem a bit scary at first, but I want you to know: you've got this. The beauty of this method is that it's not only noninvasive but also pretty dang effective. And it gives you incredible insight into your hormonal health. We're talking data that no lab test or doctor can access without you. Yes, girl, you are walking around with essential data that can help you and your doctor troubleshoot those hormone symptoms. And let's not forget, the pill is considered 99 percent effective—when used perfectly (about 91 percent with how ladies typically take it). FAM is 95 to 99 percent effective when used correctly, and it doesn't flood your body with side effects.

Let's take a quick look at four common variations of FAM and what the research says:

Fertility Basics

- Day 1 of your cycle is the first day of your period.

- Ovulation generally occurs between days 12 and 14, but this can vary, depending on your unique cycle.

- An LH surge triggers ovulation—you can use an ovulation kit to detect when you're having an LH surge.

- Your egg lives for only twenty-four hours.

- Sperm can live five or six days in your uterus.

- Your fertile window is about a max of eight days.

THE STANDARD DAYS METHOD

- This is considered the easiest method to use and requires the fewest days of abstinence or barrier contraception.

- It involves avoiding unprotected sex from day 8 through day 19 of your cycle.

- *Correct* use has resulted in a pregnancy rate of less than 5 percent a year.

- *Typical* use has resulted in a pregnancy rate of 12 percent per year.

- To be successful with this method, cycles should be 26 to 32 days.

THE CERVICAL MUCUS, OR OVULATION, METHOD

- You evaluate cervical secretions several times daily to determine your fertile days.

- It takes more time to learn this method than any other method.

- You need to abstain or use a barrier method for about fourteen to seventeen days a month.

- *Correct* use has resulted in a pregnancy rate of 3 percent.

- *Typical* use has resulted in a pregnancy rate of 23 percent.

THE TWO-DAY METHOD

- You must avoid unprotected sex on days when cervical secretions are present—all secretions are considered fertile with this method.

- You need to abstain or use a barrier method for about thirteen days each month.

- *Correct* use has resulted in a pregnancy rate of 3.5 percent.

- *Typical* use has resulted in a pregnancy rate of about 14 percent.

SYMPTOTHERMAL METHOD

- Following instructions on how to chart your cycle, you record changes in both basal body temperature (BBT) and cervical mucous (CM). The shift in hormones causes shifts in BBT.

- You need to avoid unprotected intercourse for all days with cervical secretions and four days after the last day of cervical secretions, as well as six days following a three-day spike in temperature.

- *Correct* use has resulted in a pregnancy rate of 1.8 percent; after thirteen cycles, a pregnancy rate was recorded of only 0.6 percent if there was no unprotected sex in the fertile window.

- *Typical* use has resulted in a pregnancy rate of 13 to 20 percent.

Note: Women with irregular cycles due to PCOS, breastfeeding, recent childbirth, or perimenopause are advised not to use any fertility awareness method, as pregnancy risk is higher due to the irregular nature of their cycles. Some fertility apps can help women with irregular cycles use a fertility awareness method by building in more "Don't have sex" days.

For maximum effectiveness, meet with a FAM provider who can teach you the method in detail. FAM works when women use it correctly, and a FAM educator can help ensure you've got it dialed in.

What About Fertility Apps?

Gone are the days of paper calendars and pencils. We now have technology that makes tracking your fertility—and avoiding pregnancy—even easier. That said, be sure to keep the following in mind if you use a fertility app: not all apps

Breastfeeding Does *Not* Prevent Pregnancy

Sorry, but breastfeeding is not a reliable form of birth control. Remember, ovulation comes before menstruation, which means you're fertile before you ever get your period back.

Because estrogen-containing pills reduce milk supply, the preferred pills for new moms are only progestin. While short-term and long-term studies have yet to uncover negative effects, it's possible the right outcomes aren't being measured, which is why some experts are skeptical of prescribing these hormones to a new mom. Progestin-only pills are also less effective than the combination pill.

---

are created equal. Some so-called fertility apps don't have any evidence-based methods as part of their technology.

In one study of ninety-five fertility apps, only four accurately used the Symptothermal Method and only one used the Standard Days Method. But don't let that scare you off. There are some seriously legit apps out there that can help you leverage FAM. Pair that with getting radically in tune with your body, and you've got an epic hormone-free plan to prevent pregnancy. See the Resources section for a list of apps.

Condoms

I'm guessing that you're familiar with condoms and have used one before, but if not, they are thin, stretchy pouches that cover your partner's penis during sex so his sperm is contained and blocked from finding your egg. If you're a woman who is not in a monogamous relationship (ain't no judgment here), you're going to need to use condoms even if you're using the pill or fertility awareness method. If you're not in a relationship or you're just starting to become sexually active, then a barrier method is definitely part of the conversation with your new partner, because pregnancy prevention alone is shortsighted, and there's a whole lot more than sperm that a penis can deliver to a woman. Things like HPV, HIV, and gonorrhea are no joke. So just because I'm here talking about

Can You Replace the Pill with Mathematics?

Natural Cycles was the first fertility app approved for contraception in the world! It was designed by a particle physicist and was certified as a method of birth control in the European Union in 2017. In 2018, the FDA approved this as the first contraceptive app. Natural Cycles analyzed data from more than 22,785 women and 224,563 cycles. When used correctly, this app has been shown to be 99 percent effective for contraception! Typical use results in a 93 percent efficacy rate. The app uses an algorithm to determine ovulation and your fertile window based on your basal body temperature (BBT), which you measure with a two-decimal basal thermometer first thing in the morning, before you get out of bed. Pro tip: put that thermometer by your bed and don't miss a day. That's how you bump the pregnancy prevention stat to 99 percent. A similar device and app known as Daysy has been available in the US for years, but only approved as a fertility device. That said, I have plenty of patients who have used this to avoid pregnancy with great results.

These apps take other factors into account, like sperm survival, temperature fluctuations, and cycle irregularities. The color-coded signal, alternating between red and green, makes it easy to identify your fertile days. On green days, your risk of pregnancy is low and you can have sex without protection (think: green light, go). On red days you must use protection to prevent pregnancy (think: red light, no). This app is a badass natural alternative to the birth control pill with no side effects. *Boom*, math for the win!

fertility awareness method, let's not forget the other things we ladies need to be in the know about and the other reasons to use a condom.

Just like any other form of birth control, condoms work best when used properly, so make sure you or your partner are putting them on correctly. Leaving a bit of room at the tip, the condom should be rolled down the shaft of an erect penis. After ejaculation, your partner should hold on to the rim of the condom and carefully withdraw his penis before it gets soft and the condom loosens. Condoms must be thrown away after each use; they cannot be used more than once. They should be stored in a cool, dry place—too much heat or moisture can damage them. Before use, it's also smart to check their expiration date and ensure that there aren't any tears or holes in the condom.

With correct use, condoms are 98 percent effective; when accounting for user error, they are about 82 percent effective. If you've chosen condoms solely for pregnancy prevention and you learn the fertility awareness method, you will need them only about six to eight days out of the month. Most condoms are made of latex, lambskin, or synthetic materials. Here's the lowdown on each:

- **Latex:** Some people are allergic to latex, so don't use these if you or your partner have latex sensitivities or allergies. These condoms are not compatible with oil-based lubricants or medications. They are known for providing strong protection against STIs and HIV.

- **Lambskin:** If you have a sensitivity to other condom materials, you may tolerate these better. They work with both oil- and water-based lubricants. It's important to know that lambskin condoms may not be effective at preventing STIs or HIV.

- **Synthetic:** These condoms have a low risk of allergic reaction and are compatible with both oil- and water-based lubricants (read the package to double-check the oil compatibility factor). They also have a longer shelf life. Their STI protection rate is similar to latex condoms.

Diaphragms

Diaphragms are made of flexible silicone in a dome shape that fits over your cervix—spermicide is applied to the diaphragm directly before use to create a chemical barrier to sperm. The hardest part about using a diaphragm is learning

---

Are Spermicides Safe?

While you may think spermicide will provide extra protection, spermicides in condoms have not been shown to be more effective, and in fact, spermicides actually shorten the shelf life of condoms, resulting in a higher rate of failure. Studies have also revealed a higher rate of urinary tract infection in women. My advice? Buy the condoms without spermicide!

What Is a Cervical Cap?

You may have heard of a cervical cap, and this smaller version of a diaphragm also uses spermicide and covers the cervix. In women who have not given birth, it's about 84 percent effective, but if you've already had a baby, then you have a 32 percent chance of getting pregnant using this method.

how to place it correctly, but with a little practice, you can easily master this. They are meant to be reusable and must be cleaned and stored properly. Your doctor will help you find the right fit, teach you how to properly place the diaphragm over your cervix, and discuss proper maintenance. If you have a weight change of more than 10 pounds, you will need to have your diaphragm refitted. This also applies if you have given birth or had a miscarriage or abortion. If you experience any discomfort during sex, you may also want to ensure that you have the right fit. With perfect use, diaphragms are 94 percent effective; with typical use, they are about 88 percent effective. Because they have declined in popularity since the advent of the birth control pill, they can be a bit harder to find, but you can generally find them online, and your doctor should be able to help. There is also a relatively new purple diaphragm on the market called Caya, with a more anatomically friendly design that is smaller and will fit most women. If you have a history of pelvic inflammatory disease, toxic shock syndrome, or recurrent UTIs, you may want to consider another form of birth control.

Sponge

Women like this over the diaphragm because it's easier to use and can be inserted up to twenty-four hours before sex, so you don't have to rush to the bathroom to put in your contraception just when things are heating up. The sponge is a foam disc that contains spermicide and is about two inches wide. You place it in your vagina so that it covers your cervix, blocking the entrance to your uterus, and it has a nylon loop attached for easy removal. Before you insert the sponge, you must wet it with clean water so that it's sudsy, to activate the spermicide. It starts working immediately and should be left in place for at least six hours after sex. Do not leave it in your vagina for more than thirty hours, though. The sponge should be thrown away after use. Talk to your doctor about proper insertion, placement, and use. If you've never given birth, the sponge is 88 percent effective; if you have, then it's 76 percent effective. If getting pregnant is definitely not in your life plan right now, then you may want to use a diaphragm or condom instead, paired with the fertility awareness method.

Copper IUD

This hormone-free, T-shaped device releases copper, which interferes with sperm movement and egg fertilization and may also prevent implantation of an embryo. It must be placed in your uterus by a gynecologist, midwife, or family doctor. It's more than 99 percent effective and doesn't have to be removed for seven to ten years. Because it does not contain hormones, you will have a regular menstrual cycle while using the copper IUD, although you may experience an increase in flow or menstrual cramps. The side effects are small but serious. There is a 1 percent risk of pelvic inflammatory disease within the first month and an overall risk of 0.1 percent after that. Women under the age of twenty-five have an increased risk of expulsion, where the IUD becomes dislodged and can be moved out of the body by the uterus; the average woman's risk is 3 to 5 percent. Some women have reported copper toxicity, although research is lacking in this area. But it is advised that if you have a copper storage disease, you avoid this contraceptive. If you have any concern about the copper levels, you can have them monitored with a simple blood test.

The Withdrawal Method—aka "Pulling Out"

Let's be real: we've all been there. Most women have used this method at one time or another, so if you have too, you're not alone. If you haven't used it or heard of it before, it's basically when you have sex and your partner pulls out his penis before he ejaculates—coitus interruptus. While women are often warned that this is not a reliable form of birth control, it actually works better than you think. Done correctly, it's about 96 percent effective! That said, if your man isn't quick enough, it's about 80 percent effective, so it really comes down to how much you trust your partner's agility. As a doc, I'm not comfortable with a form of contraceptive that a woman can't be 100 percent in control of. In my opinion, you're the boss of your lady parts.

Tubal Ligation

A tubal ligation is a permanent form of birth control that requires surgery, which means you should consider it only if you know you absolutely don't want children or you have already completed your family. This is commonly known as "having your tubes tied" or tubal sterilization. During this procedure, your fallopian tubes are tied or blocked, preventing an egg from traveling down or sperm from traveling up these tubes. You will still continue to have a normal menstrual cycle. It is more than 99 percent effective and potentially 100 percent irreversible— reversal requires complicated major surgery that doesn't always work.

What About the Essure Coil?

You may have heard about this other form of permanent birth control, known as the Essure coil. The Essure coil is inserted into each of your fallopian tubes, and a barrier made of scar tissue gradually forms around the inserts, which keeps sperm from reaching your eggs. What you may *not* know is that this coil can fall apart and migrate, and it's a super risky contraceptive that I don't recommend. When they fall apart, they basically explode shrapnel into your pelvis. Um, no. Trust me, there are way better options out there for you, girl.

According to Shawn Tassone, MD, PhD, Essure removal expert and board-certified ob-gyn with over twenty years of experience, "The clinical ramifications of Essure come from the PET fibers, nitinol, stainless steel, aluminum, and other endocrine-disrupting chemicals. Pelvic pain, abnormal bleeding, autoimmune disorders, rashes, joint pain, and estrogen dominance are all relatively common symptoms I see at my office with the Essure device. Finally, don't put something in your body that is permanent unless it is life or death. Essure was meant to stay and cause a permanent inflammation. Most women end up having a hysterectomy to have the device properly removed, and as such I have been advocating for women across the globe against this device [for years]."

Birth Control Options at a Glance

Method	What Is It?	Need to Know	Pros	Cons
Birth control pill	A daily pill that delivers synthetic hormones each month	With correct use, it's 99% effective; with typical use, it's 91% effective.	A simple, convenient way to prevent pregnancy	See chapters 1–10
Fertility awareness method (FAM)	A non-hormonal approach to birth control based on recognizing the signs of ovulation	You will need to abstain from sex or use additional protection during at least 8 days of your cycle. With correct use, it's 95–99% effective; with typical use, it's about 80% effective.	Noninvasive You gain incredible insight into your hormonal health. No side effects (except the chance of pregnancy, like all methods) Hormone-free	Requires abstaining from sex or use of additional protection for 8 or more days Not good for women with irregular cycles
Condom	A thin barrier of latex, lambskin, or synthetic fibers that is placed over an erect penis	With correct use, it's 98% effective; with typical use, it's about 85% effective.	Reduces the risk of UTIs and STDs Easily accessible Noninvasive Affordable	May cause an allergic reaction Reduces spontaneity Leaves room for user error Can break or tear

Birth Control Options at a Glance

Method	What Is It?	Need to Know	Pros	Cons
Diaphragm	A dome-shaped device made of flexible silicone that fits over your cervix	You must be properly fitted for a diaphragm by a doctor. With correct use, it's 94% effective; with typical use, it's 88% effective.	Affordable because it's reusable	Reduces spontaneity Leaves room for user error Needs to be refitted after weight change, birth, miscarriage, or abortion May cause discomfort during sex Not to be used if you have a history of pelvic inflammatory disease, toxic shock syndrome, or recurrent UTIs
Sponge	A wide foam disc containing spermicide that covers your cervix, blocking sperm from entering	The sponge must be left in your vagina for 6 hours after sex, but not left in your vagina for more than 30 hours. It's 88% effective if you've never given birth; it's 76% effective if you have given birth.	Easy to use Can be inserted 24 hours before sex, allowing for more spontaneity Starts working immediately Easy to remove	Not reusable and can get expensive Not as effective as other contraceptive methods

Birth Control Options at a Glance

Method	What Is It?	Need to Know	Pros	Cons
Copper IUD	A hormone-free, T-shaped device that releases copper, which interferes with sperm movement and egg fertilization	It must be placed in your uterus by a doctor. It's 99% effective and lasts 7–10 years.	Highly effective Allows for spontaneity Does not disrupt menstrual cycles Lasts a long time	May increase menstrual flow or cause cramps 1% risk of pelvic inflammatory disease in the first month; 0.1% risk after that Increases risk of expulsion in women under 25 Possible copper toxicity
Withdrawal method	A method in which your partner withdraws his penis during sex before he ejaculates sperm	With correct use, it's 96% effective; with typical use, it's 80% effective.	Allows for spontaneity Is more effective than people realize—but only when done properly	Risky—its effectiveness is in your partner's hands Not recommended if you're dead set against pregnancy Will not protect against STIs
Tubal ligation	A surgical procedure in which your fallopian tubes are tied or blocked to prevent pregnancy	This is permanent. It's more than 99% effective.	Highly effective	Requires surgery Irreversible

Key Takeaways:
Alternative Birth Control Methods

- Only you know which form of birth control is best for you, so take your time to consider your choices, do your research, and talk with your doc.

- Learning the fertility awareness method (FAM) not only can help prevent pregnancy but also is invaluable for a better understanding of your hormones.

- FAM teaches you how to recognize when you're ovulating by tracking your cervical secretions and/or basal body temperature.

- Eggs live for only twenty-four hours, but sperm can live for six days, therefore most FAM methods require abstinence or additional birth control for approximately eight days during your cycle.

- FAM is 95 to 99 percent effective when used correctly.

- Women with irregular cycles due to PCOS, breastfeeding, recent childbirth, or perimenopause are advised not to use any FAM method for pregnancy prevention.

- With correct use, condoms are 98 percent effective, and they also protect you from STIs and HIV.

- Diaphragms are reusable and must be cleaned and stored properly. They are 94 percent effective with perfect use.

- The sponge also blocks the cervix but is less reliable than the diaphragm, however it can be inserted up to twenty-four hours ahead of time.

- The copper IUD is hormone-free and lasts seven to ten years. There is a small risk of pelvic inflammatory disease.

- The withdrawal method is more effective than you think when done properly, however you must rely on your partner's skills. And girl, that is a bad idea if baby making is not on the agenda.

- Tubal ligation and the Essure coil are both permanent invasive forms of birth control.

CONCLUSION

You picked up this book because there was a voice inside you telling you that your body is capable of tremendous healing. I'd venture to guess that, like me, there was a little whisper of "Something doesn't feel quite right" when you were told about hormonal birth control. You are the one who holds the key, and now the data too, to supporting your body's incredible hormones. I know at your core you recognize you are a rebel goddess, fierce in nature with wisdom beyond this world.

I'd like to take a moment to honor you for embarking on this journey with me to take back your period and your hormones. I'm humbled and grateful that you would place your trust in me, and it is such a privilege to support you. Truly, thank you.

I see every woman in my practice like a drop of water in the pool of womanhood. When you are healed, whole, and resonating at your highest potential, you drop into that pool and create a ripple effect that touches every woman in your life. You become a living testament to what is possible, and you inspire great change. Yes, it is by healing ourselves that we will heal this crazy, radical world. Give yourself permission to heal, to be whole, to be a woman. We all need you.

As you recalibrate your hormones and create an intimate connection with your body, you will experience some incredible things. My patients who have transitioned off the pill and used the 30-Day Brighten Program tell me that for the first time they feel like themselves and it is fantastic. They get into a flow and rhythm of working with their body. Rather than feeling their body is betraying them, they can more clearly see her guiding them. They approach their body with the patience and gentleness of a mother with a small child. And the result? Period problems fade away, libidos rise just as they should, skin clears, and they find more peace. I want all of this for you and more!

Wherever you are on this journey, I want to invite you to join my community (DrBrighten.com/Community) to have support from women who have accepted the challenge of radical self-love. You are going against the social norm of hormonal suppression, and no doubt you will meet with some criticism, if you haven't already (everyone has an opinion about your body, right?). Don't worry. We've all been there. Having a community of women who get what the journey is like is an incredible resource in enabling you to heal your hormones. Speaking of resources, please take advantage of the resources I've built for you in the back of this book and at DrBrighten.com/Resources.

I have a dream of changing women's medicine for the better, but it won't happen with one doctor asking that we do better. It will happen because women like you are rejecting the old stories and writing something new about what it is to be a woman. It will happen because you are the example of what is possible, and that disrupts the stories both doctors and women have been telling themselves. And it will happen because you demand more from your health care providers . . . beyond the pill.

The change we so desperately need in women's medicine starts with you.

RECIPES

Smoothies

For the first fourteen days of the program include either Dr. Brighten Paleo Detox or Plant-Based Detox powder in your smoothies to optimize hormones. But by all means, continue using these recipes forevs!

Creamy Strawberry Cauliflower Smoothie

This smoothie tastes more like a strawberry milkshake. And while most milkshakes don't contain cruciferous vegetables, cauliflower is included here as an excellent hormone balancer. When you prep the cauliflower rice, throw some into the freezer and try incorporating it into your other smoothies too! If you're feeling "chicken," stop; this is going to get that estrogen in check *stat*.

SERVES 1

½ cup riced or finely diced cauliflower
1 cup frozen strawberries
½ cup almond milk
½ cup full-fat coconut milk*

Note: Use full-fat coconut milk from a can. Blend the top layer (the cream) with the liquid until it's completely emulsified. Save extra coconut milk in a glass jar in the fridge for up to five days. This milk can be used in smoothies or soups, or added as "creamer" for tea. Alternatively, freeze the milk to use it at another time.

¼ teaspoon ground cinnamon
1 serving detox powder or your favorite protein powder

1. Lightly steam the cauliflower.

2. In a high-speed blender, combine the steamed cauliflower, strawberries, almond milk, coconut milk, cinnamon, and detox powder. Blend the mixture until it is smooth.

Lemon Berry Boost

Some of lemon's most powerful health benefits come from limonene, which can be found mainly in the oil or peel of a lemon. For this reason, we're including lemon zest (make sure to opt for organic!). Lemon and berries happen to be a perfect match, plus the lemon will cut some of the bitterness from the kale leaves. Berries offer potent antioxidants necessary for supporting the liver and lowering inflammation. Almond butter is added too for a healthy dose of appetite-suppressing fats, keeping you calm and satisfied well into the afternoon.

SERVES 1

½ lemon
1 cup roughly chopped kale
1 cup roughly chopped spinach
1 cup frozen mixed berries
1 tablespoon almond butter
1 cup unsweetened almond milk
1 serving detox powder or your favorite protein powder
1 serving PurePaleo fiber supplement, berry flavor

1. Zest the peel of the lemon straight into a high-speed blender. Then peel the lemon with a knife and add the flesh; it's okay to include some of the white pith.

2. Add to the blender the kale, spinach, berries, almond butter, almond milk, detox powder, and fiber supplement. Blend the mixture until it is smooth.

Morning Matcha Smoothie

Replace your morning cup of coffee with this nutrient-boosting, hormone-stabilizing smoothie. Your liver will thank you for the additional boost of chlorophyll from the matcha, and you'll be saved from any caffeine crash. Plus, maca is an excellent adaptogenic herb that supports the adrenal glands.

SERVES 1

1 cup coconut water

1 cup roughly chopped spinach

1 frozen banana

1½ teaspoons matcha

1 teaspoon maca powder

¼ avocado

1 serving PurePaleo protein powder

Combine all the ingredients in a high-speed blender until the mixture is smooth.

Piña Colada Cleanser

The lime, pineapple, and coconut in this smoothie is reminiscent of a fresh piña colada! Sans alcohol, of course, and replaced with citrus zest and fresh herbs for incredible liver support.

SERVES 1

1 lime

2 romaine lettuce leaves

½ cup full-fat coconut milk (see Note, page 291)

½ cup coconut water

1 cup frozen pineapple chunks

1 tablespoon hemp seeds

1 serving PurePaleo protein powder

¼ cup fresh cilantro leaves

1. Zest the peel of the lime straight into a high-speed blender. Then peel the lime with a knife, cut the flesh into quarters, and add it to the blender.

2. Add to the blender the romaine lettuce, coconut milk, coconut water, pineapple, hemp seeds, protein powder, and cilantro, and blend the mixture until it is smooth.

Spicy Carrot Cleanser Smoothie

Pumpkin pie spice is often recognized as a seasonal ingredient, however its potent mix of cinnamon, ginger, nutmeg, allspice, and cloves make it an all-year-round anti-inflammatory blend. Fresh ginger gives this smoothie an extra digestive boost and a spicy twist, and the freshly ground black pepper is used to aid in turmeric absorption.

SERVES 1

1 medium carrot, chopped
1 cup unsweetened almond milk
1-inch piece fresh ginger, peeled and sliced
1 orange, peeled and seeded
¼ teaspoon turmeric powder
1½ teaspoons pumpkin pie spice
1 tablespoon MCT oil* or coconut oil
1 serving PurePaleo protein powder
Freshly ground black pepper to taste

Combine all the ingredients in a high-speed blender until the mixture is smooth.

---

*Note: MCTs are medium-chain triglycerides derived from coconut oil that are easily absorbed and offer a myriad of health benefits, including increased insulin sensitivity and weight loss.

Upbeet Citrus Smoothie

Put a pep in your step with this zesty, clarifying smoothie. Root vegetables like beets help to clear out estrogen and balance hormones naturally. Limonene, found in citrus peel, supports liver detoxification.

SERVES 1

½ cup chopped beets, steamed

1 navel orange, peeled

Zest from ½ lemon

1½ tablespoons fresh lemon juice

1 tablespoon MCT oil or coconut oil

¾ cup water

1 serving detox powder or your favorite protein powder

Combine all the ingredients in a high-speed blender until the mixture is smooth.

Breakfast

Brussels Sprouts Breakfast Hash

Here's a delicious way to incorporate cruciferous vegetables into your morning. Brussels sprouts are mixed with sweet potato and carrot, all of which support detoxification and provide a healthy dose of fiber for solid gut health.

SERVES 2

½ pound brussels sprouts
1 tablespoon coconut oil or ghee (once you reintroduce dairy)
½ medium sweet potato, peeled and shredded
1 medium carrot, peeled and shredded
1 teaspoon chili powder
½ teaspoon ground cumin
¼ teaspoon ground coriander
¾ teaspoon sea salt
¼ cup coarsely chopped fresh cilantro
4 large eggs

1. Trim the brussels sprouts and cut them in half. Then slice them thinly, perhaps with a mandoline. Alternatively, shred them in a food processor.

2. Heat the coconut oil in a cast-iron skillet set over medium-high heat, or a skillet you can place a lid on. Add the brussels sprouts, sweet potatoes, and carrots, and toss the vegetables to coat everything with the coconut oil. Season with the chili powder, cumin, coriander, and salt. Sprinkle with the cilantro. Stirring occasionally, allow the mixture to simmer for 5 minutes, or until the carrots and sweet potatoes are tender but not fully cooked through.

3. Make four shallow, evenly spaced indentations in the vegetable mixture, leaving enough of the mixture so you do not see the pan. Crack 1 egg into each indentation, then cover the skillet. Reduce the heat to low, and let the hash continue to simmer until the white parts of the eggs are fully cooked and the yolks are done to your preference.

Shiitake Tarragon Mini Frittatas

These mini frittatas are excellent for a grab-and-go breakfast, a protein-rich snack, or lunch over a salad. They feature shiitake mushrooms for their immune-boosting properties, alongside tarragon for a unique flavor twist. Feel free to add any other vegetables you have on hand, or switch up the spices to keep your palate interested.

MAKES 12

EQUIPMENT NEEDED: 12-CUP MUFFIN PAN

1 tablespoon coconut oil, camel hump fat, or ghee
1 cup cubed sweet potato
Sea salt and freshly ground black pepper to taste
½ cup diced onion
2½ ounces or ¾ cup sliced shiitake mushrooms
½ cup drained and chopped jarred artichoke hearts
2 cups fresh baby spinach
6 large eggs
¼ cup full-fat coconut milk (see Note, page 291)
1 teaspoon sea salt
1 teaspoon dried tarragon
½ teaspoon dried parsley
1 tablespoon avocado oil

1. Preheat the oven to 350°F.

2. Heat the coconut oil in a skillet set over medium heat. Add the sweet potatoes, and toss to coat them with the coconut oil. Add a few pinches of salt and black pepper to taste. Cover the skillet and allow the sweet potatoes to cook undisturbed for 3 minutes.

3. Uncover the skillet, add the onions, and let them cook for 1 minute, then add the mushrooms for an additional minute.

4. Add the artichoke hearts and baby spinach in the skillet, and toss all the vegetables with another pinch of salt and black pepper. Reduce the heat to low, cover the skillet again, and let the mixture simmer for 3 more minutes, or until

the spinach has wilted. Remove the vegetables from the skillet and transfer them to a bowl to slightly cool.

5. Beat the eggs in a medium bowl. Add the coconut milk, salt, tarragon, and parsley, and whisk the mixture until the coconut milk is fully combined with the eggs.

6. Brush the muffin tin with avocado oil to prevent the eggs from sticking.

7. Pour 3 tablespoons of the beaten eggs into each muffin cup, allowing room in each cup for the vegetables. Use a spoon to evenly distribute the vegetable mixture until each cup is full but not overflowing.

8. Place the muffin tin in the oven and bake for 20 minutes. Allow the mini frittatas to cool for 5 minutes before removing them from the pan and serving.

Spinach Sage Breakfast Patties

If you usually opt for eggs in the morning, try these to switch things up. Rosemary, thyme, and sage combine especially well for that classic breakfast sausage taste. Spinach is mixed in for an antioxidant boost, but you'll never taste it among the other bold flavors. These are a great emergency protein for any meal of the day, so don't forget to make an extra batch and store them in your freezer.

MAKES 8 SMALL PATTIES

2 cups loosely packed fresh baby spinach
½ pound ground beef
½ pound ground pork
¾ teaspoon sea salt
1 tablespoon minced fresh rosemary
1 teaspoon ground sage
1 teaspoon dried thyme
¼ cup coarsely chopped fresh parsley
2 tablespoons avocado oil, divided

1. Place the spinach in a food processor fitted with the S blade. Pulse until the spinach is chopped into small, coarse pieces.

2. In a bowl, combine the spinach, beef, pork, salt, rosemary, sage, thyme, and parsley. Use your hands to blend the ground meats with the spinach and spices. Form the mixture into 8 small patties.

3. Heat a cast-iron skillet set over medium-high heat and add 1 tablespoon of the avocado oil. Cook the patties in two batches, using the second tablespoon of avocado oil for the second batch. Cook each patty for 3 to 4 minutes on each side, then remove them from the heat, cover them, and allow them to rest 5 minutes before serving.

Sweet Carrot Breakfast Patties

If you naturally prefer something sweet in the morning, then these patties are for you. You'll get a nice dose of protein to keep you satisfied, combined with an excellent source of antioxidants from the carrots. Try these on top of cooked greens, or pack them with avocado to go.

MAKES 8 SMALL PATTIES

2 shallots, diced (about ¾ cup)
2 tablespoons avocado oil, divided
2 large carrots, shredded (about 1 cup)
¾ teaspoon salt, plus a pinch for the vegetables
1 pound ground chicken
½ teaspoon freshly ground black pepper
1 teaspoon ground cinnamon

1. Preheat the oven to 350°F.

2. In a skillet set over medium heat, simmer the shallots in 1 tablespoon of the avocado oil for 3 minutes.

3. Add the carrots to the skillet, plus a pinch of salt. Continue simmering the vegetables for 5 minutes, stirring occasionally.

4. Transfer the carrots and shallots to a bowl and set them aside. Allow them to cool.

5. Place the chicken in a second bowl and use your hands to gently add the ¾ teaspoon salt and the black pepper.

6. Once the carrots and shallots are cool enough to handle, add the cinnamon to them with your fingers. Then combine the carrot and shallot mixture with the chicken mixture and form it all into 8 small patties.

7. Heat a cast-iron or other oven-safe skillet over medium-high heat and add the remaining tablespoon of avocado oil. Sear the chicken patties in the skillet, 1 minute on each side. Cook them in batches if necessary.

8. Put the skillet in the oven and allow the patties to finish cooking for 10 minutes.

9. Remove the skillet from the oven and place the chicken patties on a plate to cool. Allow them to rest for 5 minutes before serving.

Snacks

Flax and Pumpkin Balls (aka Estrogen-Balancing Balls)

Here's a recipe to use when seed cycling during the first half of your menstrual cycle (or beginning on the new moon if you're on the pill). The nutrients found in flaxseeds and pumpkin seeds can help create healthy estrogen levels. Flaxseeds have lignans that can bind to estrogen, and pumpkin seeds contain fats and zinc, which are important for hormone balance. The natural sweetness of cashew butter goes great with the bitterness of ground flax and the earthy flavor of pumpkin seeds.

MAKES ABOUT 16 BALLS
SERVING SIZE: 2 BALLS
EQUIPMENT NEEDED: SPICE GRINDER, OR A BLENDER WITH A MILLING BLADE

¾ cup raw pumpkin seeds

¼ cup flaxseeds

1 tablespoon chia seeds

¼ cup toasted coconut flakes, plus extra for garnish (the garnish is optional)

½ teaspoon salt

¼ cup plus 2 tablespoons cashew butter

2 tablespoons coconut oil, melted

½ teaspoon vanilla extract

2 tablespoons honey

¼ cup currants

1. Grind the pumpkin seeds and flaxseeds in a spice grinder or a blender fitted with a milling blade.

2. In a food processor fitted with the S blade, add the ground seeds, chia seeds, coconut flakes, and salt. Pulse a few times to combine.

3. In a small bowl, stir together the cashew butter, melted coconut oil, vanilla extract, and honey. Gently mix until the cashew butter has thinned and is easily spreadable.

4. Add the cashew butter mixture to the food processor. Process the mixture until it starts to clump together or form a ball. Add the currants and pulse to combine. The ingredients should stick together when pressed between your fingers.

5. Line a glass storage container or baking dish with parchment paper. Use a tablespoon to scoop the mixture into 16 equal portions and roll each portion of the "dough" between your palms into a ball.

6. If you like, roll each ball in a small bowl of the extra coconut flakes for garnishing.

7. Place the balls in the refrigerator and allow them to set overnight.

No-Bake Sunflower Rose Cookies

These "cookies" support seed cycling during your luteal phase (or full moon to new moon if you're on the pill). Sunflower seeds contain selenium, which supports the liver and hormone regulation. Sesame seeds, like flaxseeds, contain lignans, which also support estrogen balance. Look for raw tahini (sesame seed paste); the only ingredient should be "ground sesame seeds," with no added oils. Enjoy these easy, no-bake cookies with a morning cup of tea or as a health-promoting dessert.

MAKES ABOUT 24 COOKIES

SERVING SIZE: 2 COOKIES

EQUIPMENT NEEDED: SPICE GRINDER, OR A BLENDER WITH A MILLING BLADE

1 cup raw sunflower seeds

¼ cup raw sesame seeds, plus 1 to 2 tablespoons for garnish (the garnish is optional)

½ cup toasted coconut flakes

½ teaspoon ground cardamom

1½ teaspoons ground cinnamon

¼ teaspoon sea salt

2 teaspoons lemon zest

½ cup tahini (sesame seed paste)

1 tablespoon melted coconut oil

2 tablespoons pure maple syrup

½ teaspoon vanilla extract

½ teaspoon rose water

1 to 2 tablespoons sesame seeds, for garnish (optional)

1. Grind the sunflower seeds and sesame seeds in a spice grinder or a blender fitted with a milling blade.

2. In a food processor fitted with the S blade, add the ground sunflower and sesame seeds, coconut flakes, cardamom, cinnamon, salt, and lemon zest. Pulse the mixture a few times to combine.

3. In a small bowl, stir together the tahini, melted coconut oil, maple syrup, vanilla extract, and rose water. Gently mix the combination until the tahini has thinned and is easily spreadable.

4. Add the tahini mixture to the food processor. Process well until the mixture starts to clump together or form a ball. The ingredients should stick together when pressed between your fingers.

5. Line a glass storage container or baking dish with parchment paper. Use a tablespoon to scoop the mixture into 24 uniformly sized cookies. Use your fingers to gently press each cookie into the tablespoon to create a rounded shape. Press on one side to remove them from the spoon. Sprinkle each cookie with the optional extra sesame seeds.

6. Place the cookies in the refrigerator and allow them to set overnight.

Beverages

Anti-inflammatory Turmeric Spritzer

Here's a zingy beverage that's perfect for afternoons or as a "mocktail." Lemon, ginger, and turmeric all have powerful anti-inflammatory properties and can give a nice, natural lift during a typical midday slump.

SERVES 1

1 tablespoon fresh lemon juice
½ teaspoon freshly grated ginger (use a microplane)
½ teaspoon freshly grated turmeric (use a microplane)
1 teaspoon raw honey (optional)
12 ounces sparkling water

In a large glass, gently stir the lemon juice, ginger, turmeric, and honey into the sparkling water. Alternatively, mix the lemon juice, ginger, and turmeric into 4 ounces of your favorite kombucha, then combine that with 10 ounces sparkling water.

Upgraded Golden Milk

Golden milk is a long-established beverage that's been used for centuries in Ayurvedic and traditional Chinese medicine. This recipe uses a variety of spices due to their therapeutic effects. Turmeric offers anti-inflammatory properties, coupled with black pepper for optimal absorption. Ginger is excellent for soothing the intestinal tract, and cinnamon is great for blood sugar regulation. The MCT oil is excellent for brain health and it serves as a stable energy source throughout the night. If you're looking for a quality MCT oil that also tastes good, I recommend trying Bulletproof's Brain Octane. Collagen supports gut healing and is rich in glycine, which promotes a state of calm. We also add grass-fed ghee, once dairy is reintroduced, for fat-soluble vitamins and its ability to suppress inflammation and support motility in the gut. You can whip up this Upgraded Golden Milk in minutes and enjoy it warm or over ice. Make extra in the morning and store it in a mason jar in the fridge for later or the next day.

MAKES 16 OUNCES
SERVES 2

2 cups full-fat coconut milk (see Note, page 291)
2 teaspoons turmeric powder
1 teaspoon freshly grated ginger or ginger powder
1 to 2 teaspoons MCT or coconut oil
1 to 2 teaspoons grass-fed ghee (optional)
½ teaspoon ground cinnamon
1 to 2 tablespoons grass-fed collagen hydrolysate
1 to 2 teaspoons raw honey
Pinch of freshly ground black pepper

1. Blend everything together in a high-speed blender until it is smooth.

2. Place the mixture in a small saucepan set over medium heat, and simmer for 3 to 5 minutes.

3. Drink the latte immediately or put it into the fridge for later.

Maca Latte

Maca is an adaptogenic herb that can be nourishing to the adrenal glands and beneficial for healthy estrogen and testosterone levels. Try this Maca Latte in the afternoons.

SERVES 1

¼ cup full-fat coconut milk (see Note, page 291)
¾ cup water
1 rounded teaspoon ground cinnamon
2 teaspoons maca powder
1 tablespoon collagen powder
½ pod fresh vanilla bean or ¼ teaspoon alcohol-free vanilla extract

1. Heat the coconut milk and water in a small saucepan.

2. Transfer the liquid mixture to a blender that's heat stable and add the cinnamon, maca powder, collagen powder, and vanilla. Blend the mixture until it's frothy. Serve while it's still warm.

Restorative Roots Liver Tonic

This is a great morning beverage option while you're eliminating coffee. Burdock, chicory, and dandelion combine nicely to mimic a bitter coffee-ish taste. These excellent herbs also support the liver, with the addition of milk thistle, which offers the incredible ability to protect, support, and regenerate the liver.

SERVES 1

1 teaspoon dried, ground burdock root

1 teaspoon dried, ground dandelion root

1 teaspoon dried, ground roasted chicory root

1 teaspoon ground milk thistle seeds

8 ounces water, boiling hot

2 tablespoons full-fat coconut milk (see Note, page 291)

¼ teaspoon ground cinnamon

1. Combine the burdock root, dandelion root, chicory root, and milk thistle seeds in a reusable tea bag or tea ball infuser. Steep the herbs in the hot water for 5 minutes.

2. Remove the tea bag or infuser and stir in the coconut milk and cinnamon. You may also briefly blend the mixture into a latte style if you prefer.

Salads

Baby Bok Choy Salad with Chickpea Miso

Chickpea miso is an excellent alternative to soy-based miso. You can use it to make a typical miso soup or try it as a salty component for a salad dressing, as it's used here. The creaminess of the dressing goes well with some of the more bitter, hormone- and detox-supportive vegetables, such as radicchio.

SERVES 2

3 heads baby bok choy

1 large kohlrabi

¼ head radicchio

2 large carrots, peeled and shredded

1 cup broccoli sprouts or microgreens

2 green onions, sliced

½ cup julienned fresh basil

1 tablespoon chickpea miso paste

1 tablespoon fresh lemon juice

2 tablespoons extra-virgin olive oil

1 teaspoon honey

1 clove garlic, minced

1 teaspoon freshly grated ginger

¼ cup toasted walnuts, coarsely chopped

1. Slice the bok choy on the diagonal into thin slices.

2. Peel the kohlrabi with a knife. Cut it in half, then cut it into very thin slices, using either a knife or a mandoline.

3. Cut out the core of the radicchio head, then thinly slice it.

4. Place the bok choy, kohlrabi, radicchio, carrots, sprouts or microgreens, green onions, and basil into a medium bowl and set it aside.

5. To make the dressing, in a small bowl whisk together the chickpea miso paste, lemon juice, and olive oil. (Note: Add a teaspoon or two of hot water to

the miso paste if necessary to help emulsify it.) Add the honey, garlic, and ginger. Whisk until everything is well combined, or alternatively mix it in a mini blender or food processor.

6. Toss the salad ingredients with the dressing and top it with the walnuts.

Cauliflower Tabbouleh

This rendition of tabbouleh switches out traditionally used grains for a gluten-free version with cauliflower rice, which you can make using a food processor or a box grater. Detox-supportive cauliflower will become a mainstay in your diet once you master how to make cauliflower rice! Try replacing typical rice-based dishes with the cauliflower version, such as in this recipe, or dress up cauliflower with fresh herbs and spices to serve it as a filling side dish. This dish will support your liver with a high dose of antioxidants from the cauliflower without sacrificing flavor. If you're vegetarian or vegan, add cooked chickpeas to take this from a side dish to a main meal.

SERVES 2

4 cups riced cauliflower
2 teaspoons avocado oil
3 cloves garlic, minced
1 teaspoon salt
1 teaspoon ground cumin
¼ cup currants
¼ cup toasted slivered almonds
½ cup diced carrots
¼ cup coarsely chopped fresh mint
3 green onions, thinly sliced
¼ cup coarsely chopped fresh parsley
2 tablespoons raw pumpkin or sunflower seeds, based on your cycle
1 tablespoon fresh lime juice

1. Cut the whole cauliflower into quarters and slice out the core. Break down each quarter into smaller pieces, cutting larger florets in half. To prepare the cauliflower with a food processor, fit the food processor with the S blade. Place

about a quarter of the cauliflower into the processor bowl and pulse until crumbly. Do not over-process as this will change the consistency and make your "rice" turn soft and mealy. To make the cauliflower using a box grater, after you cut the whole cauliflower into quarters and slice out the core, grate each quarter on the largest holes.

2. Set a skillet over medium heat and add the avocado oil. When the oil is hot, add the garlic and sauté it for 30 seconds.

3. Add the riced cauliflower to the skillet, add the salt, and toss the cauliflower frequently for 3 minutes, until it is tender.

4. Remove the cooked cauliflower from the skillet and place it in a medium bowl. Stir in the cumin, currants, almonds, carrots, mint, green onions, parsley, seeds, and lime juice until everything is well combined. Allow the mixture to sit 30 minutes before serving. This dish can be eaten warm or cold. Top with your favorite protein or avocado slices and enjoy.

Curly Kale with Cilantro Artichoke Pesto

Here's an ultimate detox salad. Kale and dandelion greens are some of the top detox greens that support liver and kidney function. The artichokes in this variation of pesto offer even more liver love and serve as a great replacement for the dairy and nuts used in traditional pesto.

SERVES 1

1 cup fresh cilantro leaves
1 cup fresh basil leaves
½ cup whole jarred artichoke hearts
2 cloves garlic
¼ teaspoon sea salt
Freshly ground black pepper
¼ cup extra-virgin olive oil
½ bunch curly kale, de-stemmed and finely chopped
1 cup thinly sliced dandelion greens
1 radish, thinly sliced
2 tablespoons hemp seeds
½ mango, diced (optional)

1. To make the pesto, in a food processor combine the cilantro, basil, artichoke hearts, garlic, salt, and black pepper, and process the mixture until everything is well broken down.

2. With the food processor running, drizzle in the olive oil until it's just incorporated. The pesto should be smooth and slightly runny.

3. Place the kale in a medium bowl and add half the pesto. Use your hands to massage the pesto into the greens, breaking down the tough fibers of the kale, until it looks slightly wilted. Save the rest of the pesto as a dip for vegetables or for an additional serving of the salad. (Kale is a tough green that keeps very well when made in advance, even with the dressing added.)

4. Add the dandelion greens and radish to the pesto-dressed kale.

5. This salad tastes best when allowed to sit so that the flavors can marry. When you're ready to serve it, top it with the hemp seeds and optional mango.

Liver-Cleansing Beet Slaw

Beets provide antioxidant, anti-inflammatory, and detoxification support. The raw beets in this salad, alongside the daikon, are excellent liver aids.

SERVES 2

    1 large beet, peeled and scrubbed clean
    1 medium carrot, peeled and scrubbed clean
    1 small daikon, peeled and scrubbed clean
    1 apple, cored
    Juice of 1 orange
    1 tablespoon fresh lime juice
    2 tablespoons extra-virgin olive oil
    3 teaspoons freshly grated ginger (use a microplane)
    ¼ teaspoon salt
    ¼ cup toasted walnuts
    ½ cup coarsely chopped fresh cilantro

1. Cut the beet into quarters and the carrot into three or four chunks. Place the pieces into a food processor fitted with the S blade, and process them until

they're broken down into small chunks. Transfer the mixture to a bowl and set it aside.

2. Remove the S blade and fit the food processor with a grating disc. Shred the daikon and the apple, then add them to the carrot and beet mixture.

3. To make the dressing, in a small bowl whisk together the orange juice, lime juice, olive oil, ginger, and salt. Toss the slaw with the dressing and taste, adding more lime juice, ginger, or salt to taste. Allow the slaw to sit for at least 20 minutes for the flavors to marry.

4. Serve the slaw topped with walnuts and cilantro.

Thai Zoodles with Citrus Almond Sauce

This Thai-inspired zoodle—zucchini noodle—salad is a great way to get in a variety of beneficial antioxidants via raw vegetables. The sauce is similar to a classic peanut sauce except without the soy or peanuts. Serve alongside your favorite protein.

SERVES 2

Salad
   2 large zucchini
   1 small head broccoli
   1 large carrot, peeled into long strips
   ⅛ small red cabbage, thinly sliced using a mandoline
   4 to 6 button mushrooms, sliced
   1 red bell pepper, cored and thinly sliced

Citrus Almond Sauce
   ¼ cup coconut aminos*
   ¼ cup almond butter
   2 tablespoons fresh orange juice
   1 teaspoon orange zest
   2 teaspoons fresh lime juice

*Note: Coconut aminos have a similar flavor profile to soy sauce but are made with coconut tree sap blended with mineral-rich sea salt.

2 tablespoons extra-virgin olive oil

½ teaspoon grated garlic (use a microplane)

3 teaspoons freshly grated ginger (use a microplane)

Toppings

2 green onions, thinly sliced on the diagonal

2 tablespoons toasted sesame seeds

1. Use a spiral slicer to transform the zucchini into noodles. Alternatively, use a peeler to julienne the zucchini into long strands, similar to spaghetti noodles.

2. Use a knife to cut off the top florets of the broccoli. Reserve the stalks for another use (such as in a stir-fry or a vegetable roast).

3. Combine the zucchini noodles, broccoli florets, carrots, red cabbage, mushrooms, and red bell peppers in a bowl. Set the mixture aside.

4. Place all the Citrus Almond Sauce ingredients in a blender, and briefly process them until they are just combined and creamy. Alternatively, whisk the ingredients in a bowl.

5. Pour the sauce over the zoodles and vegetables, and toss until everything is well combined.

6. Distribute the mixture between two bowls, and top each serving with the green onions and sesame seeds.

Tri-Color Cabbage Slaw

All cruciferous vegetables are excellent for supporting hormones. Napa cabbage has softer leaves than red or green cabbage, which makes it perfect for raw salads or slaws like this one.

SERVES 2

½ bunch Dinosaur or Lacinato kale, de-stemmed and julienned

½ head napa cabbage, thinly sliced

1 medium carrot, peeled and shredded

2 green onions, sliced on the diagonal

2 tablespoons extra-virgin olive oil

1 tablespoon coconut aminos

2 teaspoons dijon mustard

1 teaspoon rice vinegar

1 teaspoon honey

¼ cup toasted slivered almonds

2 tablespoons raw pumpkin or sunflower seeds, based on your cycle

1. Place the kale in a bowl by itself, then massage it with your hands to help break down its tough fibers.

2. Add the napa cabbage, carrots, and green onions to the bowl of kale. Set the vegetables aside.

3. To make the dressing, in a small bowl whisk together the olive oil, coconut aminos, dijon mustard, rice vinegar, and honey. Pour the dressing over the salad and toss until everything is evenly coated.

4. Top each serving with the almonds and seeds.

Entrées

Citrus-Marinated Flank Steak

Alongside cold-water, fatty fish, grass-fed meats can serve as a good source of omega-3 essential fatty acids. We need these types of fats to balance hormones and keep our moods stable. Try this marinade for cuts of flank, skirt, or hanger steak, which will become tender once marinated, and they'll cook quickly over high heat or on a grill. These cuts will also be some of the most economical.

SERVES 3 TO 4

3 tablespoons avocado oil

2 tablespoons coconut aminos

2 teaspoons fresh lime juice

2 teaspoons apple cider vinegar

3 cloves garlic, crushed

½ teaspoon salt

½ teaspoon ground cumin

1 teaspoon chili powder

1 teaspoon dried oregano

¼ teaspoon ground paprika

½ teaspoon dried thyme

1 pound flank steak

1. In a small bowl, whisk together the avocado oil, coconut aminos, lime juice, apple cider vinegar, garlic, salt, cumin, chili powder, oregano, paprika, and thyme.

2. Place the steak into a shallow glass or other nonreactive dish. Pour the marinade over the meat, ensuring it is entirely coated. Cover the dish and place it in the refrigerator overnight.

3. When you're ready to cook the meat, remove it from the refrigerator and let it sit at room temperature for 20 to 30 minutes.

4. Prepare a grill or warm a stove-top grill pan set over high heat. When it's hot, grill the meat 4 minutes on each side.

5. Let the meat rest on a cutting board for at least 5 minutes before slicing it against the grain.

Dijon and Almond Herb-Crusted Salmon

Here's a way to dress up salmon with minimal ingredients. Salmon provides many of the nutrients depleted by the pill, like selenium, and has the added benefit of being anti-inflammatory. While the oven's on, place some asparagus on the same sheet pan as the salmon and allow it to roast at the same time.

SERVES 2

2 6-ounce salmon fillets
Sea salt and freshly ground black pepper to taste
2 teaspoons melted coconut oil, divided
3 teaspoons dijon mustard, divided
¼ cup finely chopped almonds
2 tablespoons roughly chopped fresh parsley

1. Preheat the oven to 400°F.

2. Rinse the salmon fillets and pat them dry, then place them on a parchment-lined baking sheet.

3. Season each fillet with salt and black pepper to taste, then drizzle 1 teaspoon of the melted coconut oil over each fillet, coating both sides.

4. Spread half the mustard on the pink side of each fillet, evenly coating with the back of a spoon.

5. In a small bowl, combine the chopped almonds and parsley, then distribute the mixture evenly on top of each piece of fish.

6. Place the baking sheet in the oven and cook the salmon 10 to 12 minutes, or until it flakes with a fork. Remove the fillets from the oven and cover them with foil, allowing them to continue cooking and rest 5 minutes before serving.

Garlic Shrimp over Chili Lime Cauliflower Rice

Shrimp is an excellent source of protein, essential fatty acids, and vitamin B12—all essential nutrients to balance hormones. If you keep shrimp stashed in your freezer and prep your cauliflower in batches on a weekly basis, this single-skillet dish can be ready in 15 minutes or less.

SERVES 2

¾ pound shrimp, peeled and deveined
1 tablespoon avocado oil
2 large cloves garlic, minced
2 cups riced cauliflower*
2 teaspoons extra-virgin olive oil
2 teaspoons fresh lime juice
¾ teaspoon ground cumin
½ teaspoon chili powder
¼ teaspoon ground coriander
¼ teaspoon sea salt
Pinch of cayenne pepper
Freshly ground black pepper to taste
2 green onions, sliced
¼ cup chopped fresh cilantro
2 tablespoons toasted sunflower seeds
½ avocado, cubed

1. Rinse the shrimp and pat them dry. Set them aside.

2. Heat the avocado oil in a skillet set over medium heat. Add the garlic and stir it continuously for 30 seconds.

3. Add the shrimp to the pan and cook them for 4 minutes, tossing occasionally, until they are pink and cooked through. Remove the shrimp from the skillet and set them aside in a bowl.

4. Add the riced cauliflower to the same skillet, along with the olive oil, lime juice, cumin, chili powder, coriander, salt, cayenne pepper, and black pepper.

---

*Note: Find instructions for riced cauliflower in the Cauliflower Tabbouleh recipe on page 308.

Toss the cauliflower with the spice mixture and cook for 3 minutes, or until just tender.

5. Turn off the heat and add the shrimp, green onions, cilantro, and sunflower seeds to the skillet. Toss to combine everything well.

6. Serve topped with the avocado.

Ginger-Marinated Cod

Whitefish such as cod will be a great addition to your weekly seafood menu. Cod provides B12, folate, selenium, and magnesium, which are all depleted by the pill. Most whitefish easily lend themselves to other flavors and in many cases are cooked quickly on the stove or in the oven. Prep this the night before you plan to eat it, and pair it with bok choy, green beans, or mushrooms.

SERVES 3 TO 4

1 pound Alaskan black cod
2 tablespoons coconut aminos
2 teaspoons fresh lime juice
1 teaspoon honey
1 teaspoon freshly grated ginger (use a microplane)
2 cloves garlic, minced
½ teaspoon sesame oil
Coconut oil for cooking

1. Rinse the fish and pat it dry. Place it in a shallow glass dish and set it aside.

2. To make the marinade, whisk together the coconut aminos, lime juice, honey, ginger, garlic, and sesame oil.

3. Pour the marinade over the fish, cover the dish, and place it in the refrigerator overnight.

4. When you're ready to cook the fish, preheat the oven to 425°F.

5. Remove the marinated fish from the refrigerator when the oven is ready.

6. Heat an oven-safe skillet set over medium-high heat and melt the coconut oil. Place the fish in the skillet and sear it, skin side up, for 2 minutes or until it is golden.

7. Remove the skillet from the heat and place it in the oven, then bake the fish for 5 to 6 minutes, or until it flakes easily.

Lemongrass Thai Chicken Soup

Here's an example of a soup that will work perfectly for a variety of leftover proteins or vegetables. You can reap the benefits of anti-inflammatory garlic and ginger, gut-loving bone broth, and liver-loving greens like dandelion and kale, all without sacrificing flavor.

SERVES 2

2 teaspoons coconut oil

2 tablespoons red curry paste

2 cloves garlic, chopped

1½ tablespoons freshly grated ginger

1 shallot, diced

1 13-ounce can full-fat coconut milk (see Note, page 291)

1½ cups chicken bone broth*

2 teaspoons fish sauce

2 teaspoons coconut aminos

2 stalks lemongrass

½ cup diced carrots

2 cups roughly chopped curly kale

1 cup roughly chopped dandelion greens

6 ounces cooked shredded chicken

½ cup roughly chopped fresh cilantro leaves

½ lime, cut into wedges

Note: Look for brands that do not contain soy or sugar, such as The Flavor Chef, Bare Bones, and Kettle & Fire.

1. Heat the coconut oil in a medium pot set over medium heat. Add the red curry paste, garlic, ginger, and shallot. Toss the mixture frequently for 1 minute, then add the coconut milk, chicken bone broth, fish sauce, and coconut aminos.

2. Trim the lemongrass at both ends, 1 inch from the root and taking off any dry portions at the top. Remove the tough outer layers until you get to the tender inner part of the plant. Cut the stalks into thirds, then mash them with a meat hammer or the bottom of a mason jar. Do not over-crush the lemongrass or lemongrass strings may come apart in the soup. Add the lemongrass to the soup.

3. Turn the heat up until the soup reaches a soft boil, then reduce the heat and allow the soup to simmer for 20 minutes.

4. Add the carrots to the soup and allow the soup to cook another 5 minutes, or until the carrots are tender.

5. Remove the lemongrass and add the kale, dandelion, and shredded chicken. Cover the pot and resume simmering 3 to 5 more minutes, or until the greens have wilted and the chicken is warmed through.

6. Serve the soup topped with fresh cilantro and a squeeze from a lime wedge.

Mango Chicken Collard Wraps with Golden Curry Sauce

Collard greens serve as an excellent replacement for typical gluten-filled wraps. Plus, they're part of the cruciferous family, which means they offer similar liver- and detox-supportive nutrients that will support your hormones. The Golden Curry Sauce delivers anti-inflammatory turmeric and can be used in a variety of ways: drizzle it over cooked greens or try it as a dip for raw vegetables.

MAKES 2 WRAPS

¼ cup full-fat coconut milk (see Note, page 291)
1 teaspoon freshly grated ginger
½ teaspoon curry powder
¼ teaspoon turmeric powder
1 tablespoon fresh lemon juice

2 teaspoons extra-virgin olive oil

¼ teaspoon salt

Freshly ground black pepper to taste

3 to 4 ounces shredded chicken

2 tablespoons chopped raw cashews

¼ cup julienned fresh basil leaves

¼ cup chopped fresh mint leaves

½ mango, cut into small cubes

2 collard leaves

1. To make the sauce, whisk together the coconut milk, ginger, curry powder, turmeric powder, lemon juice, olive oil, salt, and pepper. Alternatively, place all these ingredients in a small blender or mini food processor and blend or pulse until the dressing is well combined.

2. Place the shredded chicken in a small bowl and pour half of the sauce over it, reserving the other half for dipping. Toss the chicken and sauce together with the cashews, basil, mint, and mango.

3. Prepare the collard leaves by slicing off the thick stem at the base of each leaf. Then use the butt end of a chef's knife to crush the stem along the center of the leaf.

4. Half fill a skillet with water and set it over high heat. Once the water comes to a soft boil, reduce the heat to a simmer. Add one collard green at a time and submerge it under the water for 30 seconds. The leaf should turn bright green. Remove each leaf from the water in turn and place them on a kitchen towel to pat dry.

5. Place a blanched collard leaf on a cutting board, with the base of the stem pointing toward you. Spoon half the chicken mixture in a spot three-quarters of the way down the leaf, not directly in the center. Fold up the end and each side of the leaf, on the left and right, then roll the wrap up like a burrito. Do the same with the other leaf and the remaining chicken.

6. When ready to serve, cut each collard wrap in half on a diagonal, and use the extra reserved sauce for dipping.

Mediterranean Lamb Sliders

Like wild game, lamb is a great way to change up your blood sugar–optimizing protein choices. Rosemary and basil give these sliders enough flavor to stand on their own atop a green salad, or prepare them along with a full feast of hummus, cucumbers, fermented vegetables, and Cauliflower Tabbouleh (page 308).

MAKES 6 TO 8 SLIDERS

1 pound ground lamb
¼ cup minced shallots (about two small shallots)
2 tablespoons minced fresh rosemary
2 teaspoons dried basil
1 teaspoon sea salt
Coconut oil or camel hump fat for cooking

1. In a medium bowl, combine the ground lamb, shallots, rosemary, basil, and salt using your hands, but don't overmix them.

2. Form the mixture into small patties no bigger than the size of your palm.

3. Heat the cooking oil in a skillet set over medium-high heat. Cook each patty about 3 to 4 minutes on each side, or until it's done.

4. Remove the patties from the pan and allow them to rest 5 minutes before serving.

Red Curry Salmon

Dress up your salmon with a quick coconut milk red curry sauce. The fish and sauce work well over a bed of cauliflower rice.*

SERVES 2

2 salmon fillets
¼ teaspoon salt
Freshly ground black pepper to taste

Note: See Cauliflower Tabbouleh recipe on page 308 for instructions on how to make riced cauliflower.

2 teaspoons coconut oil, camel hump fat, or ghee (once you reintroduce dairy)

1 teaspoon freshly grated ginger

1 teaspoon minced garlic

1½ teaspoons red curry paste

½ cup full-fat coconut milk (see Note, page 291)

1 teaspoon fresh lime juice

1. Preheat the oven to 350°F.

2. Rinse the salmon and pat the fillets dry on both sides. Sprinkle them with the salt and black pepper to taste. Place them in an 8 × 8-inch glass dish.

3. To prepare the red curry sauce, heat the coconut oil in a small saucepan set over medium-low heat. When the coconut oil is melted, add the ginger and garlic, and toss them in the coconut oil for 30 seconds. Add the red curry paste and continue to toss for another minute.

4. Add the coconut milk to the saucepan, stirring frequently to combine it well with the curry paste. Lower the heat, and allow the sauce to simmer for 5 minutes.

5. Remove the saucepan from the heat and stir in the lime juice. Adjust the salt and black pepper to taste.

6. Reserve a third of the red curry sauce and pour the rest over the salmon fillets in the dish. Flip each fillet to ensure the sauce evenly coats the fish.

7. Bake the fish in the oven for 12 minutes, or until the edges flake but the middle is still pink. For thicker fillets, bake them 3 to 5 minutes longer.

8. Remove the fish from the oven and cover it with foil. Allow it to rest and continue cooking for 5 minutes.

9. Serve the fillets drizzled with the reserved sauce.

Sardine Fritters

Sardines are a hormone superfood due to their B12, selenium, vitamin D, omega-3 essential fatty acid, and blood sugar–stabilizing protein content. If you haven't yet acquired a taste for sardines, try this fish cake variation—they're a food that's worth learning to love!

MAKES 5 TO 6 FISH CAKES

1 can sardines, lightly smoked and packed in extra-virgin olive oil or water
1 teaspoon lemon zest
2 teaspoons fresh lemon juice
½ cup almond flour, divided
1 tablespoon coconut flour
2 tablespoons chopped raw almonds
¼ cup coarsely chopped fresh parsley
2 tablespoons minced red onion
¼ cup chopped kalamata olives
1 clove garlic, minced
½ teaspoon freshly ground black pepper
1½ tablespoons dijon mustard
1½ teaspoons dried basil
1 teaspoon ground paprika
1 large egg
2 to 3 tablespoons avocado oil for cooking

1. Open the can of sardines and drain the water or oil, though it's fine if some of the water or oil remains. Place the fish in a large bowl, and use a fork to mash and coarsely break them apart.

2. Add the lemon zest, lemon juice, ¼ cup almond flour, and coconut flour to the bowl, and mix the flours with the fish. This will allow the flours to begin soaking up some of the moisture.

3. Add the almonds, parsley, onions, olives, garlic, black pepper, mustard, basil, and paprika, and mix until everything is well combined. Taste and adjust the seasoning to your liking. If you are still getting accustomed to the taste of sardines, you may prefer more onions, olives, garlic, or dijon, for example.

4. Whisk the egg in a small bowl and add it to the fish mixture, stirring to combine.

5. Place the extra ¼ cup almond flour in a separate small bowl and set it aside.

6. To make the patties, form ¼ cup of the fish mixture into a ball with your hands. Gently press the ball into a patty, and use your fingers to smooth along the edges where you see any cracks. Place the patty in the bowl of extra almond flour, and gently coat each side. Prepare all the patties.

7. Heat a skillet set over medium heat and add the avocado oil. When the oil is hot, set the patties in the oil and let them cook for 3½ to 4 minutes on each side, or until the almond flour coating is golden brown.

8. Place the patties on a plate lined with a paper towel to absorb any extra oil.

Seared Fish with Tomatoes and Capers

Try these Mediterranean flavors to spice up your favorite whitefish. Tomatoes are rich in antioxidants to help you replenish what was lost on the pill.

SERVES 4

4 whitefish fillets, 4 to 6 ounces each (red snapper, cod, bass, or haddock)
Sea salt and freshly ground black pepper to taste
1 teaspoon smoked paprika
1 tablespoon avocado oil
½ yellow onion, diced
2 cloves garlic, minced
1 cup chopped fresh tomatoes
2 tablespoons capers
¼ teaspoon salt
1 tablespoon coconut oil
¼ cup coarsely chopped fresh parsley
2 tablespoons sliced kalamata olives (optional)

1. Rinse the fish under cool water, then pat them dry on both sides. Season each fillet with salt, pepper, and smoked paprika on one side. Set the fish aside.

2. Warm the avocado oil in a skillet set over medium heat. Add the onions and sauté for 5 minutes, or until the onions are translucent. Add the garlic and stir frequently for another minute. Add the tomatoes, capers, and salt, and cover the skillet. Reduce the heat and simmer for 5 minutes.

3. In a separate skillet set over medium-high heat, warm the coconut oil. Lay the fish fillets seasoned-side down. Depending on the size of your skillet, you may need to cook the fish in two batches. Do not overcrowd the pan.

4. Allow the fish to cook undisturbed for 2 to 3 minutes. Each fillet should look golden and slightly crispy. While the first side is cooking, season the other side with more of the salt, pepper, and smoked paprika to taste.

5. Flip the fillets and spoon the tomato, onion, garlic, and caper mixture on top. Cover the skillet, and continue cooking the fish for 2 to 3 minutes, or until the fish flakes easily.

6. Serve the fillets topped with the parsley and optional kalamata olives.

Sesame Carrot and Cabbage Buffalo Stir-Fry

Grass-fed meats like bison and buffalo are a great way to add some variety to your protein choices. These types of meat will offer an omega-6 to omega-3 ratio that's more optimal than what you find in other meats. Buffalo is leaner than beef and therefore takes slightly less time to cook, which is perfect for this Asian-inspired stir-fry.

SERVES 4

1 tablespoon coconut oil

1 cup chopped onions

2 teaspoons freshly grated ginger

2 cloves garlic, minced

1 pound ground buffalo

¾ teaspoon sea salt

2 tablespoons coconut aminos

2 teaspoons fish sauce

2 tablespoons apple cider vinegar

¼ green cabbage, thinly sliced

2 large carrots, peeled and shredded

1 teaspoon toasted sesame oil (optional)

¼ cup sliced scallions

1 tablespoon toasted sesame seeds

1. Heat the coconut oil in a large skillet set over medium heat, then add the onions. Cook them for 5 minutes, or until they are translucent. Add the ginger and garlic and continue sautéing, stirring frequently for another minute.

2. Add the buffalo to the skillet and break apart the meat with a wooden spoon. Sprinkle the mixture with salt as the meat is browning.

3. When the meat is still pink, add the coconut aminos, fish sauce, and apple cider vinegar, tossing everything together. Also add the cabbage and carrots, then cover the pan, leaving the lid slightly ajar. Simmer the mixture for 3 to 5 minutes, or until the carrots have softened and the cabbage has shrunk.

4. Turn off the heat and remove the skillet from the burner. Add the sesame oil, if desired, and toss to coat everything.

5. When you're ready to serve, top the dish with scallions and sesame seeds.

Tikka Masala Turkey Meatballs

Indian spices like turmeric, curry, and garam masala are anti-inflammatory and add flavor. Make a double batch of this sauce to save in the freezer for a night when you're tempted to cave for takeout!

SERVES 4

Sauce

1 tablespoon coconut oil (or ghee after dairy is reintroduced)

1 onion, sliced

2 teaspoons freshly grated ginger (use a microplane)

1 teaspoon minced garlic

2 teaspoons curry powder

¼ teaspoon ground turmeric

½ teaspoon ground cinnamon

2 teaspoons ground cumin

1 teaspoon garam masala

1 teaspoon ground coriander

½ teaspoon ground ginger

½ teaspoon sea salt

1 14-ounce can tomato sauce or strained tomatoes (no sugar added)

1 cup full-fat coconut milk (see Note, page 291)

Meatballs

1 pound ground turkey meat

½ teaspoon ground coriander

1 teaspoon ground cumin

¼ teaspoon ground turmeric

2 teaspoons finely grated fresh ginger (use a microplane)

1 teaspoon salt

Coconut flour (optional)

1 tablespoon avocado oil

¼ cup coarsely chopped fresh parsley

1. First, make the sauce. Heat the coconut oil in a large skillet set over medium heat. Add the onions, and allow them to cook 5 minutes, or until they begin to turn translucent. Add the ginger and garlic, toss them with the onions for 30 seconds, then add the curry powder, turmeric, cinnamon, cumin, garam masala, coriander, ginger, and salt. Use a wooden spoon to stir in the spices. Next add the tomato sauce and coconut milk, stirring again to combine everything. Allow the sauce to simmer while you assemble the meatballs.

2. In a bowl, combine the meat with the coriander, cumin, turmeric, ginger, and salt, stirring until everything is just incorporated.

3. With your hands, roll the meat into small meatballs. If the meat is sticky and hard to work with, dust your hands with coconut flour, then roll the balls gently between your palms.

4. Heat the avocado oil in a separate skillet set over medium heat. Add the meatballs to the skillet and allow them to cook undisturbed for 4 minutes, or until the side touching the pan is browned. Flip them and cook another side for another 4 minutes.

5. While the meatballs are cooking, puree the simmering tomato sauce. Use an immersion blender in the pan or transfer the sauce to a heat-safe blender or food processor. Blend until the sauce is smooth and thick.

6. When the meatballs are browned on all sides, combine the sauce with the meatballs in one skillet, cover, and simmer everything another 7 to 10 minutes, or until the meatballs are cooked through.

7. When you're ready to serve, each portion of meatball should be topped with the sauce and a sprinkle of parsley.

Whole Chicken with Aromatics

When you cook a whole chicken, you save money, have plenty of leftovers, and are able to use the carcass in a bone broth (if you're using an organic, pasture-raised bird). This recipe uses a slow cooker, but feel free to try roasting the chicken in the oven or using other methods, such as a pressure cooker or an instant pot.

SERVES 4 TO 6

1 yellow onion, sliced
3 large carrots, peeled and cut into 2-inch pieces
1 whole chicken, 4 to 5 pounds
2 tablespoons duck fat or coconut oil
1 tablespoon sea salt
½ teaspoon freshly ground black pepper
2 leaves fresh sage
1 tablespoon fresh rosemary
½ teaspoon dried thyme

1. Line the bottom of a slow cooker with the onions and carrots.

2. Rinse the chicken and remove the giblets from its cavity. Pat the chicken dry with paper towels, then place it in the slow cooker, breast side down.

3. Melt the duck fat in a small saucepan set over medium heat. Stir in the salt, black pepper, sage, rosemary, and thyme. Pour the mixture all over the chicken in the slow cooker, then use your hands to rub the herbs and duck fat over both sides, around the legs, and under the skin.

4. Set the slow cooker to low and cook the chicken and vegetables 4 to 5 hours, or until an internal temperature reaches 165°F. Cooking time will depend on the size of the chicken.

5. When the chicken is ready, remove it from the slow cooker and let it rest for 20 minutes before carving. If you prefer crispy skin, broil the chicken for 5 to 10 minutes, until it is golden brown. Serve the chicken with the vegetables.

Zucchini Turkey Burgers

Burgers, patties, and meatballs give you a great opportunity to sneak in extra vegetables or even traditional superfoods like organ meats. Here, shredded zucchini serves as a bulking agent in the burgers, a perfect alternative to typical flour or breadcrumbs. Serve in a lettuce wrap or on a bed of greens.

MAKES 6 TO 8 PATTIES

1 small zucchini

1 pound ground turkey

¼ cup minced shallots

¼ cup coarsely chopped fresh parsley

1 teaspoon dried oregano

½ teaspoon garlic powder

1 teaspoon sea salt

½ teaspoon lemon zest

1 tablespoon avocado oil

1. Shred the zucchini and place it in a fine-mesh bag (such as one used for making nut milks) or in a few layers of cheesecloth. Set this in a colander in the sink to let the zucchini release some water.

2. In a medium bowl, combine the turkey meat, shallots, parsley, oregano, garlic powder, salt, and lemon zest, mixing briefly.

3. Return to the zucchini, and squeeze out any excess water. Then add the zucchini to the turkey mixture and combine.

4. Form the mixture into small patties, and use your fingers to smooth the edges.

5. Heat a skillet set over medium-high heat, and add the avocado oil. Cook the turkey patties 3 minutes on each side, or until they are lightly browned and cooked through.

QUIZ ANSWERS FOR THE MENSTRUAL CYCLE MYTH BUSTER

1. False
2. False
3. False
4. True
5. False
6. False
7. False
8. False
9. True
10. False

APPENDIX 2

YOUR FIFTH VITAL SIGN

Menstrual Cycle Symptoms	What It Could Be Telling You
Heavy or long period	Iron deficiency anemia, estrogen dominance, thyroid disease, fibroids or polyps, certain cancers, copper IUD, endometriosis, bleeding disorder
Painful period	Infection, endometriosis, fibroids, ovarian cysts, increased inflammation, elevated prostaglandins
Light period	Low estrogen (possibly from eating a low-fat diet, overexercising, low body weight), PBCS, POI, or perimenopause
Late or irregular period	Pregnancy, stress (physical or mental trauma), PCOS, perimenopause, postpartum, diabetes, celiac disease, thyroid disease
Short cycle	No ovulation or corpus luteum didn't form correctly, elevated prolactin, obesity, endometriosis, PCOS, thyroid disease, anorexia, POI, perimenopause
Missing period	PBCS, functional hypothalamic amenorrhea, pituitary dysfunction, ovarian dysfunction, hypothyroidism, pregnancy, menopause, POI
Mid-cycle pain	Ovulation (usually benign, but discuss any new onset of pain with your doctor)
Premenstrual spotting	Fibroids, infection, endometriosis, cancer, pregnancy
Pain or bleeding with sex	Low estrogen, infection, anatomy/position, endometriosis, ovarian cysts, fibroids, cancer
PMS	Hormonal imbalance (estrogen dominance, low progesterone), nutrient depletion

RESOURCES

Dr. Jolene Brighten Online

Connect with me online to stay in the know on hormones, birth control, and lady-centered inspiration.

Website: www.drbrighten.com

The Dr. Brighten Community: www.drbrighten.com/community

Instagram: @drjolenebrighten

Facebook: https://www.facebook.com/drbrighten

Pinterest: https://www.pinterest.com/drjolenebrighten

YouTube: https://www.youtube.com/c/jbrightennaturopathicdoctor

Twitter: https://twitter.com/drbrighten

Dr. Brighten's Programs, Books, and Recommended Supplements

You can find these resources on my website, www.drbrighten.com:

Post–Birth Control Hormone Reset program

Clear Skin Rx Master Class

Cosmic Cycle Sync program

Hormone Revolution Detox program

Post–Birth Control Rx program

The Fertility Master Class

The Libido Master Class

The PMS Master Class

The Postpartum Master Class

The Thyroid Master Class

Healing Your Body Naturally After Childbirth (book)

Dr. Brighten supplements can be found at www.drbrighten.com/supplements.

Finding a Naturopathic or Functional Medicine Doctor

Naturopathic and functional medicine physicians are excellent at identifying a root cause and helping you heal your hormones naturally. Only a licensed practitioner can order and interpret labs, diagnose, and prescribe treatment. Make sure you're working with a licensed provider such as an NMD/ND, MD, DC, PharmD, LAc, NP, PA, PT, or RD to best address your health care needs. A nutrition practitioner, massage therapist, counselor, personal trainer, or health coach can be a great addition to your health care team, but they are not trained to diagnose or treat health issues, including PBCS. Make sure your practitioner graduated from an accredited college and holds a license if they use the words "diagnose," "treat," or "prescribe." I'm all for a collaborative team, but I want to make sure you get the best care by meeting with a licensed practitioner.

RUBUS HEALTH

The 30-Day Brighten Program practitioners at my women's medicine clinic work with women all over the world who are struggling with post–birth control syndrome, looking to come off birth control, and wanting to heal from common hormone conditions like thyroid disease, adrenal dysfunction, PMS, and more.

329 NE Couch Street
Portland, Oregon 97232
(503) 498-8830
info@rubushealth.com
For more information, visit https://drbrighten.com/work-with-me.
Women's health consulting is available for virtual clients.

AMERICAN ASSOCIATION OF NATUROPATHIC PHYSICIANS (AANP)
NATIONAL DIRECTORY OF NATUROPATHIC PHYSICIANS

https://www.naturopathic.org/AF_MemberDirectory.asp?version=2

THE INSTITUTE FOR FUNCTIONAL MEDICINE (IFM)
FUNCTIONAL MEDICINE PRACTITIONER DIRECTORY
https://www.ifm.org/find-a-practitioner

Lab Testing

For a list of recommended labs, including where and how to get tested, please visit www.drbrighten.com/resources.

Beyond the Pill Downloadable Guides and Resources

I've put together tons of helpful resources, guides, and e-books to support you during your healing journey. You can find all of them at www.beyondthepill book.com.

Beyond the Pill Grocery Shopping List
A weekly shopping list for all the meals in the 30-day plan.

Hormone Support Recipe Guide
A guide to the best sources of organic, non-GMO foods online that support hormone health, plus shopping tips, recipes, and more.

Beyond the Pill Supplement Guide
A list of all the supplements recommended in the 30-day plan with links to purchase them.

Birth Control Nutrient Depletions Food Guide
A tool to help you leverage your diet to replenish nutrient stores both on and off birth control.

Hormone-Friendly Kitchen Swaps
A resource to help you make healthy swaps in the kitchen that leave you feeling satisfied.

Nontoxic Household Cleaning Products Guide
A guide to removing hormone-harming toxins from your environment to improve your hormone health.

Green Beauty Guide
A list of the products I use and recommend to my patients to help you avoid exposing yourself to hormone-disrupting chemicals.

Beyond the Pill Lab Guide
A list of recommended labs to holistically assess your health on or off birth control, along with links to how to order your own tests.

Hormone-Friendly Cooking and Kitchen Supplies and Equipment

For a list of hormone-safe pots, pans, and utensils, visit https://drbrighten.com/hormone-friendly-kitchen.

Femtech Apps and Devices

Clue app

Dame—women-designed pleasure devices

Daysy fertility monitor

Dot app

Kindara app

Lioness orgasm-tracking device and app

My Moontime app

Natural Cycles fertility monitor

Tia app—the first data-driven app to provide personalized birth control advice

Books

The Adrenal Reset Diet by Dr. Alan Christianson

Beautiful You by Nat Kringoudis

Clean Skin from Within by Dr. Trevor Cates

Code Red by Lisa Lister

Cooking for Hormone Balance by Magdalena Wszelaki

8 Steps to Reverse Your PCOS by Dr. Fiona McCulloch

Hashimoto's Protocol by Dr. Izabella Wentz

Healing PCOS by Amy Medling

Honoring Our Cycles by Katie Singer

The Hormone Cure by Dr. Sara Gottfried

A Mind of Your Own by Dr. Kelly Brogan

Period Repair Manual by Dr. Lara Briden

Periods Gone Public by Jennifer Weiss-Wolf

The Pill: Are You Sure It's for You? by Jane Bennett and Alexandra Pope

Sweetening the Pill by Holly Grigg-Spall

Taking Charge of Your Fertility by Toni Weschler

Wild Feminine by Tami Lynn Kent

Woman Code by Alisa Vitti

Women's Bodies, Women's Wisdom by Dr. Christiane Northrup

Fertility Awareness Method

Association of Fertility Awareness Professionals, http://www
.fertilityawarenessprofessionals.com

FertilityUK, http://www.fertilityuk.org

The Fifth Vital Sign, http://www.5thvitalsign.com

For a full list of FAM educators, please visit www.drbrighten
.com/resources.

REFERENCES

Chapter 1: Real Talk About the Pill

Akinloye, O., et al. "Effects of Contraceptives on Serum Trace Elements, Calcium, and Phosphorus Levels." *West Indian Medical Journal* 60, no. 3 (2011): 308–15.

Aminzadeh, A., et al. "Frequency of Candidiasis and Colonization of *Candida Albicans* in Relation to Oral Contraceptive Pills." *Iran Red Crescent Medical Journal* 18, no. 10 (2016): e38909.

Anderson, K. E., et al. "Effects of Oral Contraceptives on Vitamin Metabolism." *Advances in Clinical Chemistry* 18 (1976): 247–87.

Bird, S. T., et al. "Irritable Bowel Syndrome and Drospirenone-Containing Oral Contraceptives; A Comparative-Safety Study." *Current Drug Safety* 7, no. 1 (2012): 8–15.

Centers for Disease Control, Cancer and Steroid Hormone Study. "Long-Term Oral Contraceptive Use and the Risk of Breast Cancer." *JAMA* 249, no. 12 (1983): 1591–95.

Chasan-Taber, L., et al. "Epidemiology of Oral-Contraceptives and Cardiovascular Disease." *Annals of Internal Medicine* 128, no. 6 (1998): 467–77.

Davidson, N. E., et al. "Good News About Oral Contraceptives." *New England Journal of Medicine* 346 (2002): 2078–79.

Davis, A. R., et al. "Occurrence of Menses or Pregnancy After Cessation of a Continuous Oral Contraceptive." *Fertility and Sterility* 89, no. 5 (2008): 1059–63.

Dreon, D. M., et al. "Oral Contraceptive Use and Increased Plasma Concentration of C-Reactive Protein." *Life Sciences* 73, no. 10 (2003): 1245–52.

Dunn, N., et al. "Oral Contraceptives and Myocardial Infarction: Results of the MICA Case Control Study." *British Medical Journal* 318, no. 7198 (1999): 1579–84.

Fröhlich, M., et al. "Oral Contraceptive Use Is Associated with a Systemic Acute Phase Response." *Fibrinolysis and Proteolysis* 13, no. 6 (1999): 239–44.

Gingnell, M., et al. "Oral Contraceptive Use Changes Brain Activity and Mood in Women with Previous Negative Affect on the Pill—A Double-Blinded, Placebo-Controlled Randomized Trial of a Levonorgestrel-Containing Combined Oral Contraceptive." *Psychoneuroendocrinology* 38, no. 7 (2013): 1133–44.

Hankinson, S. E., et al. "A Prospective Study of Oral Contraceptive Use and Risk of Breast Cancer (Nurses' Health Study, United States)." *Cancer Causes and Control* 8, no. 1 (1997): 65–72.

Hertel, J., et al. "Evidence for Stress-Like Alterations in the HPA-Axis in Women Taking Oral Contraceptives." *Scientific Reports* 7, no. 1 (2017): 14111.

Hickman, R. J., et al. "C-Reactive Protein Is Elevated in Atypical but Not Nonatypical Depression: Data from the National Health and Nutrition Examination Survey (NHANES) 1999–2004." *Journal of Behavioral Medicine* 37, no. 4 (2014): 621–29.

Hock, H. "The Pill and the College Attainment of American Women and Men." Florida State University, September 15, 2005. http://paa2006.princeton.edu/papers/61745.

Jenkins, T. A., et al. "Influence of Tryptophan and Serotonin on Mood and Cognition with a Possible Role of the Gut-Brain Axis." *Nutrients* 8, no. 1 (2016): 56.

Khalili, H., et al. "Oral Contraceptives, Reproductive Factors, and Risk of Inflammatory Bowel Disease." *Gut* 62, no. 8 (2013): 1153–59.

Khier, L. A. M. "Effects of Oral Contraceptives on the Thyroid Function in Sudanese Females." BS thesis, University of Khartoum, 2000. https://core.ac.uk/download /pdf/71669046.pdf.

Kluft, C., et al. "Pro-Inflammatory Effects of Oestrogens During Use of Oral Contraceptives and Hormone Replacement Treatment." *Vascular Pharmacology* 39, no. 3 (2002): 149–54.

Kulkami, J. "Depression as a Side Effect of the Contraceptive Pill." *Expert Opinion on Drug Safety* 6, no. 4 (2007): 371–74.

Lewis, M. A., et al. "Third Generation Oral Contraceptives and Risk of Myocardial Infarction: An International Case-Control Study. Transnational Research Group on Oral Contraceptives and the Health of Young Women." *British Medical Journal* 312, no. 7023 (1996): 88–90.

Maes, M., et al. "The Effects of Glucocorticoids on the Availability of L-Tryptophan and Tyrosine in the Plasma of Depressed Patients." *Journal of Affective Disorders* 18, no. 2 (1990): 121–27.

Marchbanks, P. A., et al. "Oral Contraceptives and the Risk of Breast Cancer." *New England Journal of Medicine* 346 (2002): 2025–32.

Montoya, E. R., et al. "How Oral Contraceptives Impact Social-Emotional Behavior and Brain Function." *Trends in Cognitive Sciences* 21, no. 2 (2017): 125–36.

Myint, A. M., et al. "The Role of the Kynurenine Metabolism in Major Depression." *Journal of Neural Transmission* 119, no. 2 (2012): 245–51.

Nassaralla, C. L., et al. "Characteristics of the Menstrual Cycle after Discontinuation of Oral Contraceptives." *Journal of Women's Health* 20, no. 2 (2011): 169–77.

Oinonen, K. A., et al. "To What Extent Do Oral Contraceptives Influence Mood and Affect?" *Journal of Affective Disorders* 70, no. 3 (2002): 229–40.

"Oral-Contraceptive Use and the Risk of Breast Cancer. The Cancer and Steroid Hormone Study of the Centers for Disease Control and the National Institute of Child Health and Human Development." *New England Journal of Medicine* 315, no. 7 (1986): 405–11.

Palan, P. R., et al. "Effects of Oral, Vaginal, and Transdermal Hormonal Contraception on Serum Levels of Coenzyme Q10, Vitamin E, and Total Antioxidant Activity." *Obstetrics and Gynaecology International* (2010), article ID 925635.

Palmery, M., et al. "Oral Contraceptives and Changes in Nutritional Requirements." *European Review for Medical and Pharmacological Sciences* 17, no. 13 (2013): 1804–13.

Panzer, C., et al. "Impact of Oral Contraceptives on Sex Hormone–Binding Globulin and Androgen Levels: A Retrospective Study in Women with Sexual Dysfunction." *Journal of Sexual Medicine* 3, no. 1 (2006): 104–13.

Park, B., et al. "Oral Contraceptive Use, Micronutrient Deficiency, and Obesity Among Premenopausal Females in Korea: The Necessity of Dietary Supplements and Food Intake Improvement." *PLoS One* 11, no. 6 (2016): e0158177.

Peddie, B. A., et al. "Relationship Between Contraceptive Method and Vaginal Flora." *Australian and New Zealand Journal of Obstetrics and Gynaecology* 24, no. 3 (1984): 217–18.

Piltonen, T., et al. "Oral, Transdermal, and Vaginal Combined Contraceptives Induce an Increase in Markers of Chronic Inflammation and Impair Insulin Sensitivity in Young

Healthy Normal-Weight Women: A Randomized Study." *Human Reproduction* 27, no. 10 (2012): 3046–56.

Rohr, U. D. "The Impact of Testosterone Imbalance on Depression and Women's Health." *Maturitas* 41, suppl. 1 (2002): S25–46.

Sarkar, M., et al. "Influence of Moonlight on the Birth of Male and Female Babies." *Nepal Medical College Journal* 7, no. 1 (2005): 62–64.

Schatz, D. L., et al. "Effects of Oral Contraceptives and Pregnancy on Thyroid Function." *Canadian Medical Association Journal* 99, no. 18 (1968): 882–86.

Schmidt, A., et al. "Oral Contraceptive Use and Vaginal Candida Colonization." [In German.] *Zentralbl Gynakol* 119, no. 11 (1997): 545–49.

Shakerinejad, G., et al. "Factors Predicting Mood Changes in Oral Contraceptive Pill Users." *Reproductive Health* 10 (2013): 45.

Skovlund, C. W., et al. "Association of Hormonal Contraception with Depression." *JAMA Psychiatry* 73, no. 11 (2016): 1154–62.

Skovlund, C. W., et al. "Association of Hormonal Contraception with Suicide Attempts and Suicides." *American Journal of Psychiatry* 175, no. 4 (2017): 336–42.

Smith, J. S., et al. "Cervical Cancer and Use of Hormonal Contraceptives: A Systematic Review." *Lancet* 361, no. 9364 (2003): 1159–67.

Sorgdrager, F. J. H., et al. "The Association Between the Hypothalamic Pituitary Adrenal Axis and Tryptophan Metabolism in Persons with Recurrent Major Depressive Disorder and Healthy Controls." *Journal of Affective Disorders* 222 (2017): 32–39.

Talukdar, N., et al. "Effect of Long-Term Combined Oral Contraceptive Pill Use on Endometrial Thickness." *Obstetrics and Gynecology* 120, no. 2, pt. 1 (2012): 348–54.

Turna, B., et al. "Women with Low Libido: Correlation of Decreased Androgen Levels with Female Sexual Function Index." *International Journal of Impotence Research* 17, no. 2 (2005): 148–53.

Van de Wijgert, J. H., et al. "Hormonal Contraception Decreases Bacterial Vaginosis but Oral Contraception May Increase Candidiasis: Implications for HIV Transmission." *AIDS* 27, no. 13 (2013): 2141–53.

Van Hylckama Vlieg, A., et al. "The Venous Thrombotic Risk of Oral Contraceptives, Effects of Oestrogen Dose and Progestogen Type: Results of the MEGA Case-Control Study." *British Medical Journal* 339 (2009): b2921.

Vessey, M., et al. "Oral Contraceptive Use and Cancer. Findings in a Large Cohort Study, 1968–2004." *British Journal of Cancer* 95, no. 3 (2006): 385–89.

Wang, Q., et al. "Effects of Hormonal Contraception on Systemic Metabolism: Cross-Sectional and Longitudinal Evidence." *International Journal of Epidemiology* 45, no. 5 (2016): 1445–57.

Webb, J. L. "Nutritional Effects of Oral Contraceptive Use: A Review." *Journal of Reproductive Medicine* 25, no. 4 (1980): 150–56.

White, T., et al. "Effects of Transdermal and Oral Contraceptives on Estrogen-Sensitive Hepatic Proteins." *Contraception* 74, no. 4 (2006): 293–96.

Wiegratz, I., et al. "Effect of Four Different Oral Contraceptives on Various Sex Hormones and Serum-Binding Globulins." *Contraception* 67, no. 1 (2003): 25–32.

Zimmerman, Y., et al. "The Effect of Combined Oral Contraception on Testosterone Levels in Healthy Women: A Systematic Review and Meta-Analysis." *Human Reproduction Update* 20, no. 1 (2014): 76–105.

Chapter 2: The Lowdown on Your Hormones

Adams, J. M., et al. "The Midcycle Gonadotropin Surge in Normal Women Occurs in the Face of an Unchanging Gonadotropin-Releasing Hormone Pulse Frequency." *Journal of Clinical Endocrinology and Metabolism* 79 (1994): 858–64.

Baerwald, A. R., O. A. Olatunbosun, and R. A. Pierson. "Ovarian Follicular Development Is Initiated During the Hormone-Free Interval of Oral Contraceptive Use." *Contraception* 70, no. 5 (2004): 371–77.

Cella, F., G. Giordano, and R. Cordera. "Serum Leptin Concentrations During the Menstrual Cycle in Normal-Weight Women: Effects of an Oral Triphasic Estrogen-Progestin Medication." *European Journal of Endocrinology* 142 (2000): 174–78.

Clubb, E. "Natural Methods of Family Planning." *Journal of the Royal Society of Health* 106, no. 4 (1986): 121–26.

Crosignani, P. G., et al. "Ovarian Activity During Regular Oral Contraceptive Use." *Contraception* 54, no. 5 (1996): 271–73.

Davis, A. R., et al. "Occurrence of Menses or Pregnancy After Cessation of a Continuous Oral Contraceptive." *Fertility and Sterility* 89, no. 5 (2008): 1059–63.

Filicori, M., J. P. Butler, and W. F. Crowley Jr. "Neuroendocrine Regulation of the Corpus Luteum in the Human. Evidence for Pulsatile Progesterone Secretion." *Journal of Clinical Investigation* 73 (1984): 1638–47.

Filicori, M., N. Santoro, G. R. Merriam, and W. F. Crowley Jr. "Characterization of the Physiological Pattern of Episodic Gonadotropin Secretion Throughout the Human Menstrual Cycle." *Journal of Clinical Endocrinology and Metabolism* 62 (1986): 1136–44.

Fleischer, A. C., G. C. Kalemeris, and S. S. Entman. "Sonographic Depiction of the Endometrium During Normal Cycles." *Ultrasound in Medicine and Biology* 12, no. 4 (1986): 271–77.

Gipson, I. K., et al. "The Amount of MUC5B Mucin in Cervical Mucus Peaks at Midcycle." *Journal of Clinical Endocrinology and Metabolism* 86, no. 2 (2001): 594–600.

Gougeon, A. "Dynamics of Follicular Growth in the Human: A Model from Preliminary Results." *Human Reproduction* 1, no. 2 (1986): 81–87.

Gougeon A. "Dynamics of Human Follicular Growth: A Morphologic Perspective." In *The Ovary*, edited by E. Y. Adashi and P. C. K. Leung (New York: Raven Press, 1993), p. 21.

Hall, J. E., D. A. Schoenfeld, K. A. Martin, and W. F. Crowley Jr. "Hypothalamic Gonadotropin-Releasing Hormone Secretion and Follicle-Stimulating Hormone Dynamics During the Luteal-Follicular Transition." *Journal of Clinical Endocrinology and Metabolism* 74, no. 3 (1992): 600–7.

Jacobs, H. S., et al. "Post-'Pill' Amenorrhoea—Cause or Coincidence?" *British Medical Journal* 2, no. 6092 (1977): 940–42.

Kenealy, B. P., et al. "Neuroestradiol in the Hypothalamus Contributes to the Regulation of Gonadotropin Releasing Hormone Release." *Journal of Neuroscience* 33, no. 49 (2013): 19051–9.

Knochenhauer, E., and R. Azziz. "Ovarian Hormones and Adrenal Androgens During a Woman's Life Span." *Journal of the American Academy of Dermatology* 45, suppl. 3 (2001): S105–15.

Milsom, I., and T. Korver. "Ovulation Incidence with Oral Contraceptives: A Literature Review." *BMJ Sexual and Reproductive Health* 34, no. 4 (2008): 237–46.

Morrison, A. I. "Persistence of Spermatozoa in the Vagina and Cervix." *British Journal of Venereal Diseases* 48, no. 2 (1972): 141–43.

Richards, J. S. "Hormonal Control of Gene Expression in the Ovary." *Endocrine Reviews* 15, no. 6 (1994): 725–51.

Rosenberg, M. J., and M. S. Waugh. "Oral Contraceptive Discontinuation: A Prospective Evaluation of Frequency and Reasons." *American Journal of Obstetrics and Gynecology* 179, no. 3 (1998): 577–82.

Sherman, B. M., and S. G. Korenman. "Hormonal Characteristics of the Human Menstrual Cycle Throughout Reproductive Life." *Journal of Clinical Investigation* 55, no. 4 (1975): 699–706.

Simpson, E. R. "Sources of Estrogen and Their Importance." *Journal of Steroid Biochemistry and Molecular Biology* 86, no. 3–5 (2003): 225–30.

Stocco, C., C. Telleria, and G. Gibori. "The Molecular Control of Corpus Luteum Formation, Function, and Regression." *Endocrine Reviews* 28 (2007): 117–49.

Suarez, S. S., and A. A. Pacey. "Sperm Transport in the Female Reproductive Tract." *Human Reproduction Update* 12, no. 1 (2006): 23–37.

Taylor, A. E., et al. "Midcycle Levels of Sex Steroids Are Sufficient to Recreate the Follicle-Stimulating Hormone but Not the Luteinizing Hormone Midcycle Surge: Evidence for the Contribution of Other Ovarian Factors to the Surge in Normal Women." *Journal of Clinical Endocrinology and Metabolism* 80, no. 5 (1995): 1541–47.

Treloar, A. E., R. E. Boynton, B. G. Behn, and B. W. Brown. "Variation of the Human Menstrual Cycle Through Reproductive Life." *International Journal of Fertility* 12, no. 1, pt. 2 (1967): 77–126.

Tsafriri, A., S. Y. Chun, and R. Reich. "Follicular Rupture and Ovulation." In *The Ovary*, edited by E. Y. Adashi and P. C. K. Leung (New York: Raven Press, 1993), p. 227.

Welt, C. K., et al. "Frequency Modulation of Follicle-Stimulating Hormone (FSH) During the Luteal-Follicular Transition: Evidence for FSH Control of Inhibin B in Normal Women." *Journal of Clinical Endocrinology and Metabolism* 82, no. 8 (1997): 2645–52.

Chapter 3: Post–Birth Control Syndrome

Behre, H. M., et al. "Efficacy and Safety of an Injectable Combination Hormonal Contraceptive for Men." *Journal of Clinical Endocrinology and Metabolism* 101, no. 12 (2016): 4779–88.

Boyle, N. B., et al. "The Effects of Magnesium Supplementation on Subjective Anxiety and Stress—A Systematic Review." *Nutrients* 9, no. 5 (2017): 429.

Cormia, F. E. "Alopecia from Oral Contraceptives." *JAMA* 201, no. 8 (1967): 635–37.

Darney, P. D. "OC Practice Guidelines: Minimizing Side Effects." *International Journal of Fertility and Women's Medicine,* Supplement 1 (1997): 158–69.

Gebel Berg, E. "The Chemistry of the Pill." *ACS Central Science* 1, no. 1 (2015): 5–7.

Gnoth, C., et al. "Cycle Characteristics After Discontinuation of Oral Contraceptives." *Gynecological Endocrinology* 16, no. 4 (2002): 307–17.

Griffiths, W. A. D. "Diffuse Hair Loss and Oral Contraceptives." *British Journal of Dermatology* 88, no. 1 (1973): 31–36.

Gröber, U., et al. "Magnesium in Prevention and Therapy." *Nutrients* 7, no. 9 (2015): 8199–226.

Hammond, N., et al. "Nutritional Neuropathies." *Neurologic Clinics* 31, no. 2 (2013): 477–89.

Johnson, S. "The Multifaceted and Widespread Pathology of Magnesium Deficiency." *Medical Hypotheses* 56, no. 2 (2001): 163–70.

Kia, A. S., et al. "The Association Between the Risk of Premenstrual Syndrome and Vitamin D, Calcium, and Magnesium Status Among University Students: A Case Control Study." *Health Promotion Perspectives* 5, no. 3 (2015): 225–30.

Lawrence, J. "NSAID Use May Prevent Fertile Women from Ovulating." *Pharmaceutical Journal* 294, no. 7868/9 (2015).

McDowell, L. R. *Vitamins in Animal and Human Nutrition.* 2nd ed. Ames: Iowa State University Press, 2000, 265–310.

McIntosh, E. N. "Treatment of Women with the Galactorrhea-Amenorrhea Syndrome with Pyridoxine (Vitamin B6)." *Journal of Clinical Endocrinology and Metabolism* 42, no. 6 (1976): 1192–95.

Mehta, K., et al. "Antiproliferative Effect of Curcumin (Diferuloylmethane) Against Human Breast Tumor Cell Line." *Anticancer Drugs* 8, no. 5 (1997): 470–81.

Mendonça, L. L. F., et al. "Non-Steroidal Anti-Inflammatory Drugs as a Possible Cause for Reversible Infertility." *Rheumatology* 39, no. 8 (2000): 880–82.

Nabel, Elizabeth, G. "Coronary Heart Disease in Women—an Ounce of Prevention." *New England Journal of Medicine* 343 (August 24, 2000): 572–74.

Parry, B. L., et al. "Oral Contraceptives and Depressive Symptomatology: Biologic Mechanisms." *Comprehensive Psychiatry* 20, no. 4 (1979): 347–58.

Piltonen, T., et al. "Oral, Transdermal, and Vaginal Combined Contraceptives Induce an Increase in Markers of Chronic Inflammation and Impair Insulin Sensitivity in Young Healthy Normal-Weight Women: A Randomized Study." *Human Reproduction* 27, no. 10 (2012): 3046–56.

Posaci, C., et al. "Plasma Copper, Zinc, and Magnesium Levels in Patients with Premenstrual Tension Syndrome." *Acta Obstetricia et Gynecologica Scandinavica* 73, no. 6 (1994): 452–55.

Practice Committee of American Society for Reproductive Medicine. "Current Evaluation of Amenorrhea." *Fertility and Sterility* 90, suppl. 5 (2008): S219–25.

Rajizadeh, A., et al. "Effect of Magnesium Supplementation on Depression Status in Depressed Patients with Magnesium Deficiency: A Randomized, Double-Blind, Placebo-Controlled Trial." *Nutrition* 35 (2017): 56–60.

Ramos, P. M., et al. "Female Pattern Hair Loss: A Clinical and Pathophysiological Review." *Anais Brasileiros de Dermatologia* 90, no. 4 (2015): 529–43.

Roberts, S. C., et al. "MHC-Correlated Odour Preferences in Humans and the Use of Oral Contraceptives." *Proceedings of the Royal Society B: Biological Sciences* 275, no. 1652 (2008): 2715–22.

Rojas-Walsson, R., et al. "Diagnosis and Management of Post-Pill Amenorrhea." *Journal of Family Practice* 13, no. 2 (1981): 165–69.

Stewart, M. E., et al. "Effect of Cyproterone Acetate-Ethinyl Estradiol Treatment on the Proportions of Linoleic and Sebaleic Acids in Various Skin Surface Lipid Classes." *Archives of Dermatological Research* 278, no. 6 (1986): 481–85.

University of Pennsylvania School of Medicine. "Two Thirds of Women Interested in Stopping Their Periods but Unsure About Safety." *ScienceDaily.* October 4, 2007.

Wang, J. G., et al. "The Complex Relationship Between Hypothalamic Amenorrhea and Polycystic Ovary Syndrome." *Journal of Clinical Endocrinology and Metabolism* 93, no. 4 (2008): 1394–97.

Wedekind, C., et al. "Body Odour Preferences in Men and Women: Do They Aim for Specific MHC Combinations or Simply Heterozygosity?" *Proceedings of the Royal Society B: Biological Sciences* 264, no. 1387 (1997): 1471–79.

Wyatt, K. M., et al. "Efficacy of Vitamin B-6 in the Treatment of Premenstrual Syndrome: Systematic Review." *British Medical Journal* 318, no. 7195 (1999): 1375–81.

Chapter 4: Take Back Your Period

American College of Obstetricians and Gynecologists Committee on Adolescent Health Care. "Menstruation in Girls and Adolescents: Using the Menstrual Cycle as a Vital Sign." Committee Opinion No. 651. December 2015. https://www.acog.org/-/media/Committee -Opinions/Committee-on-Adolescent-Health-Care/co651.pdf?dmc=1&ts=20180626 T0442019822.

Bouchard, C., et al. "Use of Oral Contraceptive Pills and Vulvar Vestibulitis: A Case-Control Study." *American Journal of Epidemiology* 156, no. 3 (2002): 254–61.

Goldstein, A. T., et al. "Polymorphisms of the Androgen Receptor Gene and Hormonal Contraceptive Induced Provoked Vestibulodynia." *Journal of Sexual Medicine* 11, no. 11 (2014): 2764–71.

Chapter 5: Birth Control Hormone Detox 101

Baum, J. K., et al. "Possible Association Between Benign Hepatomas and Oral Contraceptives." *Lancet* 2, no. 7835 (1973): 926–29.

Blackwell. "Oral Contraceptive Pill May Prevent More than Pregnancy: Could Cause Long-Term Problems with Testosterone." ScienceDaily.com. January 5, 2006. https:// www.sciencedaily.com/releases/2006/01/060104232338.htm.

Bradlow, H. L., et al. "Multifunctional Aspects of the Action of Indole-3-Carbinol as an Antitumor Agent." *Annals of the New York Academy of Sciences* 889 (1999): 204–13.

Bradlow, H. L., et al. "2-Hydroxyestrone: The 'Good' Estrogen." *Journal of Endocrinology* 150 suppl. (1996): S259–65.

Dawling, S., et al. "Catechol-o-methyltransferase (COMT)-Mediated Metabolism of Catechol Estrogens: Comparison of Wild-Type and Variant COMT Isoforms." *Cancer Research* 61, no. 18 (2001): 6716–22.

De Waal, E. J., et al. "Differential Effects of 2,3,7,8-Tetrachlorodibenzo-p-dioxin, Bis(tri-n butyltin) Oxide, and Cyclosporine on Thymus Histophysiology." *Critical Reviews in Toxicology* 27, no. 4 (1997): 381–430.

Edmondson, H. A., et al. "Liver-Cell Adenomas Associated with Use of Oral Contraceptives." *New England Journal of Medicine* 294, no. 9 (1976): 470–72.

Eramo, S., et al. "Estrogenicity of Bisphenol A Released from Sealants and Composites: A Review of the Literature." *Annali di Stomatologia* 1, nos. 3–4 (2010): 14–21.

Etminan, M., et al. "Oral Contraceptives and the Risk of Gallbladder Disease: A Comparative Safety Study." *Canadian Medical Association Journal* 183, no. 8 (2011): 899–904.

Fishman, W. H. "Beta-Glucuronidase in the Metabolic Conjugation of Estrogenic Hormones." *Federation Proceedings* 6, no. 1, pt. 2 (1947): 251.

Fowke, J. H., et al. "*Brassica* Vegetable Consumption Shifts Estrogen Metabolism in Healthy Postmenopausal Women." *Cancer Epidemiology, Biomarkers and Prevention* 9, no. 8 (2000): 773–79.

Hofmann, A. F. "The Continuing Importance of Bile Acids in Liver and Intestinal Disease." *Archives of Internal Medicine* 159, no. 22 (1999): 2647–58.

Klatskin, G. "Hepatic Tumors: Possible Relationship to Use of Oral Contraceptives." *Gastroenterology* 73, no. 2 (1977): 386–94.

Kuhl, H., et al. "The Effect of Sex Steroids and Hormonal Contraceptives upon Thymus and Spleen on Intact Female Rats." *Contraception* 28, no. 6 (1983): 587–601.

Lord, R. S., et al. "Estrogen Metabolism and the Diet-Cancer Connection: Rationale for Assessing the Ratio of Urinary Hydroxylated Estrogen Metabolites." *Alternative Medicine Review* 7, no. 2 (2002): 112–29.

Marciani, L., et al. "Effects of Various Food Ingredients on Gallbladder Emptying." *European Journal of Clinical Nutrition* 67, no. 11 (2013): 1182–87.

McCann, S. E., et al. "Changes in 2-Hydroxyestrone and 16α-Hydroxyestrone Metabolism with Flaxseed Consumption: Modification by COMT and CYP1B1 Genotype." *Cancer Epidemiology, Biomarkers and Prevention* 16, no. 2 (2007): 256–62.

Meissner, K. "Hemorrhage Caused by Ruptured Liver Cell Adenoma Following Long-Term Oral Contraceptives: A Case Report." *Hepatogastroenterology* 45, no. 19 (1998): 224–25.

Mueck, A. O., et al. "Estradiol Metabolism and Malignant Disease." *Maturitas* 43, no. 1 (2002): 1–10.

Muti, P., et al. "Estrogen Metabolism and Risk of Breast Cancer: A Prospective Study of the 2:16Alpha-Hydroxyestrone Ratio in Premenopausal and Postmenopausal Women." *Epidemiology* 11, no. 6 (2000): 635–40.

Nime, F., et al. "The Histology of Liver Tumors in Oral Contraceptive Users Observed During a National Survey by the American College of Surgeons Commission on Cancer." *Cancer* 44, no. 4 (1979): 1481–89.

Nwachukwu, J. C., et al. "Resveratrol Modulates the Inflammatory Response Via an Estrogen Receptor-Signal Integration Network." *Elife* 3 (2014): e02057.

Oredipe, O. A., et al. "Dietary Glucarate-Mediated Inhibition of Initiation of Diethyl-nitrosamine-Induced Hepatocarcinogenesis." *Toxicology* 74, nos. 2–3 (1992): 209–22.

Panzer, C., et al. "Impact of Oral Contraceptives on Sex Hormone–Binding Globulin and Androgen Levels: A Retrospective Study in Women with Sexual Dysfunction." *Journal of Sexual Medicine* 3, no. 1: 104–13.

Prigge, J. R., et al. "Hepatocyte DNA Replication in Growing Liver Requires Either Glutathione or a Single Allele of *TXNRD1*." *Free Radical Biology and Medicine* 52, no. 4 (2012): 803–10.

Racine, A., et al. "Menopausal Hormone Therapy and Risk of Cholecystectomy: A Prospective Study Based on the French E3N Cohort." *Canadian Medical Association Journal* 185, no. 7 (2013): 555–61.

Raftogianis, R., et al. "Estrogen Metabolism by Conjugation." *Journal of the National Cancer Institute Monographs* 27 (2000): 113–24.

Rooks, J. B., et al. "Epidemiology of Hepatocellular Adenoma. The Role of Oral Contraceptive Use." *JAMA* 242, no. 7 (1979): 644–48.

Rosenberg, L. "The Risk of Liver Neoplasia in Relation to Combined Oral Contraceptive Use." *Contraception* 43, no. 6 (1991): 643–52.

Ruotolo, R., et al. "Anti-Estrogenic Activity of a Human Resveratrol Metabolite." *Nutrition, Metabolism, and Cardiovascular Diseases* 23, no. 11 (2013): 1086–92.

Schuurman, H. J., et al. "Chemicals Trophic for the Thymus: Risk for Immunodeficiency and Autoimmunity." *International Journal of Immunopharmacology* 14, no. 3 (1992): 369–75.

Shortell, C. K., et al. "Hepatic Adenoma and Focal Nodular Hyperplasia." *Surgery, Gynecology, and Obstetrics* 173, no. 5 (1991): 426–31.

Søe, K. L., et al. "Liver Pathology Associated with the Use of Anabolic-Androgenic Steroids." *Liver* 12, no. 2 (1992): 73–79.

Steiner, J. L., et al. "Dose-Dependent Benefits of Quercetin on Tumorigenesis in the C3(1)/SV40Tag Transgenic Mouse Model of Breast Cancer." *Cancer Biology and Therapy* 15, no. 11 (2014): 1456–67.

Ursin, G., et al. "Urinary 2-Hydroxyestrone/16Alpha-Hydroxyestrone Ratio and Risk of Breast Cancer in Postmenopausal Women." *Journal of the National Cancer Institute* 91, no. 12 (1999): 1067–72.

Wallwiener, C. W., et al. "Prevalence of Sexual Dysfunction and Impact of Contraception in Female German Medical Students." *Journal of Sexual Medicine* 7, no. 6 (2010): 2139–48.

Wu, Z., et al. "Progesterone Inhibits L-Type Calcium Currents in Gall Bladder Smooth Muscle Cells." *Journal of Gastroenterology and Hepatology* 25, no. 12 (2010): 1838–43.

Yang, C. Z., et al. "Most Plastic Products Release Estrogenic Chemicals: A Potential Health Problem That Can Be Solved." *Environmental Health Perspectives* 119, no. 7 (2011): 989–96.

Zhang, L. Q., et al. "Potential Therapeutic Targets for the Primary Gallbladder Carcinoma: Estrogen Receptors." *Asian Pacific Journal of Cancer Prevention* 14, no. 4 (2013): 2185–90.

Chapter 6: Gut Check

Adlercreutz, H., et al. "Studies on the Role of Intestinal Bacteria in Metabolism of Synthetic and Natural Steroid Hormones." *Journal of Steroid Biochemistry* 20, no. 1 (1984): 217–29.

Ashrafi, M., et al. "The Presence of Anti-Thyroid and Anti-Ovarian Auto-Antibodies in Familial Premature Ovarian Failure." *International Journal of Fertility and Sterility* 1, no. 4 (2008): 171–74.

Bast, A., et al. "Celiac Disease and Reproductive Health." *Practical Gastroenterology* 35, no. 10 (2009): 10–21.

Bernier, M. O., et al. "Combined Oral Contraceptive Use and the Risk of Systemic Lupus Erythematosus." *Arthritis and Rheumatology* 61, no. 4 (2009): 476–81.

Blasi, F., et al. "The Effect of N-Acetylcysteine on Biofilms: Implications for the Treatment of Respiratory Tract Infections." *Respiratory Medicine* 117 (2016): 190–97.

Bultman, S. J. "The Microbiome and Its Potential as a Cancer Preventive Intervention." *Seminars in Oncology* 43, no. 1 (2016): 97–106.

Campana, R., et al. "Strain-Specific Probiotic Properties of Lactic Acid Bacteria and Their Interference with Human Intestinal Pathogens Invasion." *Gut Pathogens* 9, no. 1 (2017): 12.

Cook, L. C., et al. "The Role of Estrogen Signaling in a Mouse Model of Inflammatory Bowel Disease: A Helicobacter Hepaticus Model." *PLoS One* 9, no. 4 (2014): e94209.

Costenbader, K. H., et al. "Reproductive and Menopausal Factors and Risk of Systemic Lupus Erythematosus in Women." *Arthritis and Rheumatology* 56, no. 4 (2007): 1251–62.

Crujeiras, A. B., et al. "Leptin Resistance in Obesity: An Epigenetic Landscape." *Life Sciences* 140 (2015): 57–63.

Cummings, S. R., et al. "Prevention of Breast Cancer in Postmenopausal Women: Approaches to Estimating and Reducing Risk." *Journal of the National Cancer Institute* 101, no. 6 (2009): 384–98.

Cutolo, M., et al. "Estrogens and Autoimmune Diseases." *Annals of the New York Academy of Sciences* 1089 (2006): 538–47.

Dai, Z. L., et al. "L-Glutamine Regulates Amino Acid Utilization by Intestinal Bacteria." *Amino Acids* 45, no. 3 (2013): 501–12.

De Kort, S., et al. "Leaky Gut and Diabetes Mellitus: What Is the Link?" *Obesity Reviews* 12, no. 6 (2011): 449–58.

Dinicola, S., et al. "N-Acetylcysteine as Powerful Molecule to Destroy Bacterial Biofilms. A Systematic Review." *European Review for Medical and Pharmacological Sciences* 18, no. 19 (2014): 2942–48.

Divi, R. L., et al. "Anti-thyroid Isoflavones from Soybean: Isolation, Characterization, and Mechanisms of Action." *Biochemical Pharmacology* 54, no. 10 (1997): 1087–96.

Engen, P. A., et al. "The Gastrointestinal Microbiome: Alcohol Effects on the Composition of Intestinal Microbiota." *Alcohol Research* 37, no. 2 (2015): 223–36.

Faculty of Family Planning and Reproductive Health Care Clinical Effectiveness Unit. "Contraceptive Choices for Women with Inflammatory Bowel Disease." *Journal of Family Planning and Reproductive Health Care* 29, no. 3 (2003): 127–35. http://srh.bmj.com /content/familyplanning/29/3/127.full.pdf.

Fasano, A. "Leaky Gut and Autoimmune Diseases." *Clinical Reviews in Allergy and Immunology* 42, no. 1 (2012): 71–78.

Fasano, A. "Zonulin, Regulation of Tight Junctions, and Autoimmune Diseases." *Annals of the New York Academy of Sciences* 1258 (2012): 25–33.

Fasano, A., et al. "Mechanisms of Disease: The Role of Intestinal Barrier Function in the Pathogenesis of Gastrointestinal Autoimmune Diseases." *Nature Clinical Practice. Gastroenterology and Hepatology* 2, no. 9 (2005): 416–22.

Fénichel, P., et al. "Prevalence, Specificity, and Significance of Ovarian Antibodies During Spontaneous Premature Ovarian Failure." *Human Reproduction* 12, no. 12 (1997): 2623–28.

Fresko, I., et al. "Anti-*Saccharomyces Cerevisiae* Antibodies (ASCA) in Behçet's Syndrome." *Clinical and Experimental Rheumatology* 23, suppl. 38 (2005): S67–70.

Frieri, M. "Neuroimmunology and Inflammation: Implications for Therapy of Allergic and Autoimmune Diseases." *Annals of Allergy, Asthma, and Immunology* 90, no. 6, suppl. 3 (2003): S34–40.

Gameiro, C. M., et al. "Menopause and Aging: Changes in the Immune System—A Review." *Maturitas* 67, no. 4 (2010): 316–20.

Gawron, L., et al. "Oral Contraceptive Use and Crohn's Disease Complications." *Gastroenterology* 151, no. 5 (2016): 1038–39. https://www.gastrojournal.org/article/S0016-5085 (16)35072-7/fulltext?code=ygast-site.

Harmon, Q. E., et al. "Use of Estrogen-Containing Contraception Is Associated with Increased Concentrations of 25-Hydroxy Vitamin D." *Journal of Clinical Endocrinology and Metabolism* 101, no. 9 (2016): 3370–77. https://academic.oup.com/jcem/article /101/9/3370/2806637.

Hartmann, P., et al. "The Intestinal Microbiome and the Leaky Gut as Therapeutic Targets in Alcoholic Liver Disease." *Frontiers in Physiology* 3 (2012): 402.

Hu, M. L., et al. "Effect of Ginger on Gastric Motility and Symptoms of Functional Dyspepsia." *World Journal of Gastroenterology* 17, no. 1 (2011): 105–10.

Jacobsen, B. K., et al. "Soy Isoflavone Intake and the Likelihood of Ever Becoming a Mother: The Adventist Health Study-2." *International Journal of Women's Health* 6 (2014): 377–84.

Jefferson, W. N. "Adult Ovarian Function Can Be Affected by High Levels of Soy." *Journal of Nutrition* 140, no. 12 (2010): 2322S–25S.

Jessop, D. S., et al. "Effects of Stress on Inflammatory Autoimmune Disease: Destructive or Protective?" *Stress* 7, no. 4 (2004): 261–66.

Khalili, H., et al. "Association Between Long-Term Oral Contraceptive Use and Risk of Crohn's Disease Complications in a Nationwide Study." *Gastroenterology* 150, no. 7 (2016): 1561–67.e1. https://www.gastrojournal.org/article/S00165085(16)002328/full text?code=ygast-site.

Khalili, H., et al. "Oral Contraceptives, Reproductive Factors, and Risk of Inflammatory Bowel Disease." *Gut* 62, no. 8 (2013): 1153–59.

Khalili, H., et al. "Sa1231 Oral Contraceptive Use and Risk of Surgery Among Crohn's Patients." *Gastroenterology* 148, no. 4, suppl. 1 (2015): S264–65.

Kwa, M., et al. "The Intestinal Microbiome and Estrogen Receptor–Positive Female Breast Cancer." *Journal of the National Cancer Institute* 108, no. 8 (2016).

Long, M. D., et al. "Shifting Away from Estrogen-Containing Oral Contraceptives in Crohn's Disease." *Gastroenterology* 150, no. 7 (2016): 1518–20.

Luborsky, J., et al. "Ovarian Antibodies, FSH, and Inhibin B: Independent Markers Associated with Unexplained Infertility." *Human Reproduction* 15, no. 5 (2000): 1046–51.

Lyngsø, J., et al. "Association Between Coffee or Caffeine Consumption and Fecundity and Fertility: A Systematic Review and Dose-Response Meta-Analysis." *Clinical Epidemiology* 9 (2017): 699–719.

McMichael-Phillips, D. F., et al. "Effects of Soy-Protein Supplementation on Epithelial Proliferation in the Histologically Normal Human Breast." *American Journal of Clinical Nutrition* 68, suppl. 6 (1998): 1431S–35.

Mulak, A., et al. "Sex Hormones in the Modulation of Irritable Bowel Syndrome." *World Journal of Gastroenterology* 20, no. 10 (2014): 2433–48.

Petrakis, N. L., et al. "Stimulatory Influence of Soy Protein Isolate on Breast Secretion in Pre- and Postmenopausal Women." *Cancer Epidemiology, Biomarkers, and Prevention* 5, no. 10 (1996): 785–94.

Petri, M., et al. "Oral Contraceptives and Systemic Lupus Erythematosus." *Arthritis and Rheumatism* 40, no. 5 (1997): 797–803.

Quah, S. Y., et al. "N-Acetylcysteine Inhibits Growth and Eradicates Biofilm of Enterococcus Faecalis." *Journal of Endodontics* 38, no. 1 (2012): 81–85.

Rizzo, G., et al. "Soy, Soy Foods, and Their Role in Vegetarian Diets." *Nutrients* 10, no. 1 (2018): 43.

Schliep, K. C., et al. "Caffeinated Beverage Intake and Reproductive Hormones Among Premenopausal Women in the BioCycle Study." *American Journal of Clinical Nutrition* 95, no. 2 (2012): 488–97.

Scrimgeour, A. G., et al. "Zinc and Micronutrient Combinations to Combat Gastrointestinal Inflammation." *Current Opinion in Clinical Nutrition and Metabolic Care* 12, no. 6 (2009): 653–60.

Sieron, D., et al. "The Effect of Chronic Estrogen Application on Bile and Gallstone Composition in Women with Cholelithiasis." *Minerva Endocrinologica* 41, no. 1 (2016): 19–27.

Skrovanek, S., et al. "Zinc and Gastrointestinal Disease." *World Journal of Gastrointestinal Pathophysiology* 5, no. 4 (2014): 496–513.

Stojanovich, L., et al. "Stress as a Trigger of Autoimmune Disease." *Autoimmunity Reviews* 7, no. 3 (2008): 209–13.

Sun, C. L., et al. "Dietary Soy and Increased Risk of Bladder Cancer: The Singapore Chinese Health Study." *Cancer Epidemiology, Biomarkers, and Prevention* 11, no. 12 (2002): 1674–77.

Tayyebi-Khosroshahi, H., et al. "Effect of Treatment with Omega-3 Fatty Acids on C-Reactive Protein and Tumor Necrosis Factor-Alfa in Hemodialysis Patients." *Saudi Journal of Kidney Diseases and Transplantation* 23, no. 3 (2012): 500–6.

Thorne Research. "Calcium-D-Glucarate." *Alternative Medicine Review* 7, no. 4 (2002): 336–39.

Tralau, T., et al. "Insights on the Human Microbiome and Its Xenobiotic Metabolism: What Is Known About Its Effects on Human Physiology?" *Expert Opinion on Drug Metabolism and Toxicology* 11, no. 3 (2015): 411–25.

Tung, K. S. K., et al. "Mechanisms of Autoimmune Disease in the Testis and Ovary." *Human Reproduction Update* 1, no. 1 (1995): 35–50.

Vojdani, A., et al. "Cross-Reaction Between Gliadin and Different Food and Tissue Antigens." *Food and Nutrition Sciences* 4, no. 1 (2013): 20–32.

Wang, B., et al. "Glutamine and Intestinal Barrier Function." *Amino Acids* 47, no. 10 (2015): 2143–54.

Wang, B., et al. "L-Glutamine Enhances Tight Junction Integrity by Activating CaMK Kinase 2-AMP-Activated Protein Kinase Signaling in Intestinal Porcine Epithelial Cells." *Journal of Nutrition* 146, no. 3 (2016): 501–8.

White, L., et al. "Prevalence of Dementia in Older Japanese-American Men in Hawaii: The Honolulu-Asia Aging Study." *JAMA* 276, no. 12 (1996): 955–60.

Williams, W. V. "Hormonal Contraception and the Development of Autoimmunity: A Review of the Literature." *Linacre Quarterly* 84, no. 3 (2017): 275–95.

Yan, Y., et al. "Omega-3 Fatty Acids Prevent Inflammation and Metabolic Disorder Through Inhibition of NLRP3 Inflammasome Activation." *Immunity* 38, no. 6 (2013): 1154–63.

Yang, Q., et al. "Added Sugar Intake and Cardiovascular Diseases Mortality Among US Adults." *JAMA Internal Medicine* 174, no. 4 (2014): 516–24. https://jamanetwork.com/journals/jamainternalmedicine/fullarticle/1819573.

Yates, C. M., et al. "Pharmacology and Therapeutics of Omega-3 Polyunsaturated Fatty Acids in Chronic Inflammatory Disease." *Pharmacology and Therapeutics* 141, no. 3 (2014): 272–82.

Zhao, T., et al. "N-Acetylcysteine Inhibit Biofilms Produced by Pseudomonas Aeruginosa." *BMC Microbiology* 10 (2010): 140.

Zitvogel, L. "Cancer and the Gut Microbiota: An Unexpected Link." *Science Translational Medicine* 7, no. 271 (2015): 271ps1.

Chapter 7: Energize Your Thyroid and Adrenals

Ajjan, R., et al. "The Pathogenesis of Hashimoto's Thyroiditis: Further Developments in Our Understanding." *Hormone and Metabolic Research* 47, no. 10 (2015): 702–10.

Alexander, E. K., et al. "2017 Guidelines of the American Thyroid Association for the Diagnosis and Management of Thyroid Disease During Pregnancy and the Postpartum." *Thyroid* 27, no. 3 (2017): 315–89. https://www.liebertpub.com/doi/pdfplus/10.1089/thy.2016.0457.

Ayhan, M. G., et al. "The Prevalence of Depression and Anxiety Disorders in Patients with Euthyroid Hashimoto's Thyroiditis: A Comparative Study." *General Hospital Psychiatry* 36, no. 1 (2014): 95–98.

Bajaj, J. K., et al. "Various Possible Toxicants Involved in Thyroid Dysfunction: A Review." *Journal of Clinical and Diagnostic Research* 10, no. 1 (2016): FE01–03.

Balázs, C., et al. "Effect of Selenium on HLA-DR Expression of Thyrocytes." *Autoimmune Diseases* (2012), article ID 374635.

Ban, Y., et al. "The Contribution of Immune Regulatory and Thyroid Specific Genes to the Etiology of Graves' and Hashimoto's Diseases." *Autoimmunity* 36, nos. 6–7 (2003): 367–79.

Ban, Y., et al. "Genetic Susceptibility in Thyroid Autoimmunity." *Pediatric Endocrinology Reviews* 3, no. 1 (2005): 20–32.

Blackwell, J. "Evaluation and Treatment of Hyperthyroidism and Hypothyroidism." *Journal of the American Association of Nurse Practitioners* 16, no. 10 (2004): 422–25.

Bunevicius, A., et al. "Hypothalamic-Pituitary-Thyroid Axis Function in Women with a Menstrually Related Mood Disorder." *Psychosomatic Medicine* 74, no. 8 (2012): 810–16.

Caturegli, P., et al. "Autoimmune Thyroid Diseases." *Current Opinion in Rheumatology* 19, no. 1 (2007): 44–48.

Cauci, S., et al. "Effects of Third-Generation Oral Contraceptives on High-Sensitivity C-Reactive Protein and Homocysteine in Young Women." *Obstetrics and Gynecology* 111, no. 4 (2008): 857–64.

Chandrasekhar, K., et al. "A Prospective, Randomized Double-Blind, Placebo-Controlled Study of Safety and Efficacy of a High-Concentration Full-Spectrum Extract of *Ashwagandha* Root in Reducing Stress and Anxiety in Adults." *Indian Journal of Psychological Medicine* 34, no. 3 (2012): 255–62.

Checchi, S., et al. "L-Thyroxine Requirement in Patients with Autoimmune Hypothyroidism and Parietal Cell Antibodies." *Journal of Clinical Endocrinology and Metabolism* 93, no. 2 (2008): 465–69.

Ch'ng, C. L., et al. "Celiac Disease and Autoimmune Thyroid Disease." *Clinical Medicine and Research* 5, no. 3 (2007): 184–92.

Cojocaru, M., et al. "Multiple Autoimmune Syndrome." *Maedica* 5, no. 2 (2010): 132–34.

Contempre, B., et al. "Effect of Selenium Supplementation in Hypothyroid Subjects of an Iodine and Selenium Deficient Area: The Possible Danger of Indiscriminate Supplementation of Iodine-Deficient Subjects with Selenium." *Journal of Endocrinology and Metabolism* 73, no. 1 (1991): 213–15.

Daher, R., et al. "Consequences of Dysthyroidism on the Digestive Tract and Viscera." *World Journal of Gastroenterology* 15, no. 23 (2009): 2834–38.

Dama, M., et al. "Thyroid Peroxidase Autoantibodies and Perinatal Depression Risk: A Systematic Review." *Journal of Affective Disorders* 198 (2016): 108–21.

De Herder, W. W., et al. "On the Enterohepatic Cycle of Triiodothyronine in Rats; Importance of the Intestinal Microflora." *Life Sciences* 45, no. 9 (1989): 849–56.

Divani, A. A., et al. "Effect of Oral and Vaginal Hormonal Contraceptives on Inflammatory Blood Biomarkers." *Mediators of Inflammation* (2015), article ID 379501.

Drutel, A., et al. "Selenium and the Thyroid Gland: More Good News for Clinicians." *Clinical Endocrinology* 78, no. 2 (2013): 155–64.

Ebert, E. C. "The Thyroid and the Gut." *Journal of Clinical Gastroenterology* 44, no. 6 (2010): 402–6.

Fan, Y., et al. "Selenium Supplementation for Autoimmune Thyroiditis: A Systematic Review and Meta-Analysis." *International Journal of Endocrinology* 2014, no. 1 (2014): 904573.

Fisher, D. A. "Physiological Variations in Thyroid Hormones: Physiological and Pathophysiological Considerations." *Clinical Chemistry* 42, no. 1 (1996): 135–39.

Foster, J. A., et al. "Stress and the Gut-Brain Axis: Regulation by the Microbiome." *Neurobiology of Stress* 7 (2017): 124–36.

Gärtner, R., et al. "Selenium Supplementation in Patients with Autoimmune Thyroiditis Decreases Thyroid Peroxidase Antibodies Concentrations." *Journal of Clinical Endocrinology and Metabolism* 87, no. 4 (2002): 1687–91.

Gerenova, J., et al. "Clinical Significance of Autoantibodies to Parietal Cells in Patients with Autoimmune Thyroid Diseases." *Folia Medica* 55, no. 2 (2013): 26–32.

Gierach, M., et al. "Hashimoto's Thyroiditis and Carbohydrate Metabolism Disorders in Patients Hospitalised in the Department of Endocrinology and Diabetology of Ludwik Rydygier Collegium Medicum in Bydgoszcz Between 2001 and 2010." *Endokrynologia Polska* 63, no. 1 (2012): 14–17.

Girdler, S. S., et al. "Historical Sexual Abuse and Current Thyroid Axis Profiles in Women with Premenstrual Dysphoric Disorder." *Psychosomatic Medicine* 66, no. 3 (2004): 403–10.

Gorini, P., et al. "The Stimulating Effect of a Cytosol Extract from Regenerating Liver on Isolated Hepatocytes and the Positive Role of Insulin." *Italian Journal of Surgical Sciences* 18, no. 3 (1988): 201–5.

Guggenheim, A. G., et al. "Immune Modulation from Five Major Mushrooms: Application to Integrative Oncology." *Integrative Medicine* 13, no. 1 (2014): 32–44.

Guhad, F. A., et al. "Salivary IgA as a Marker of Social Stress in Rats." *Neuroscience Letters* 216, no. 2 (1996): 137–40.

Haugen, B. R. "Drugs that Suppress TSH or Cause Central Hypothyroidism." *Best Practice and Research. Clinical Endocrinology and Metabolism* 23, no. 6 (2009): 793–800.

Hays, M. T. "Thyroid Hormone and the Gut." *Endocrine Research* 14, nos. 2–3 (1988): 203–24.

Houston, M. "The Role of Magnesium in Hypertension and Cardiovascular Disease." *Journal of Clinical Hypertension* 13, no. 11 (2011): 843–47.

Hybenova, M., et al. "The Role of Environmental Factors in Autoimmune Thyroiditis." *Neuro Endocrinology Letters* 31, no. 3 (2010): 283–89.

Kahaly, G. J., et al. "Thyroid Hormone Action in the Heart." *Endocrine Reviews* 26, no. 5 (2005): 704–28. https://academic.oup.com/edrv/article/26/5/704/2355198.

Kakuno, Y., et al. "Menstrual Disturbances in Various Thyroid Diseases." *Endocrine Journal* 57, no. 12 (2010): 1017–22.

Keely, E. J. "Postpartum Thyroiditis: An Autoimmune Thyroid Disorder Which Predicts Future Thyroid Health." *Obstetric Medicine* 4, no. 1 (2011): 7–11.

Kimura, H., et al. "Chemokine Orchestration of Autoimmune Thyroiditis." *Thyroid* 17, no. 10 (2007): 1005–11.

Kloosterboer, H. J., et al. "Effects of Three Low-Dose Contraceptive Combinations on Sex Hormone–Binding Globulin, Corticosteroid Binding Globulin and Antithrombin III Activity in Healthy Women: Two Monophasic Desogestrel Combinations (Containing 0.020 or 0.030 mg Ethinylestradiol) and One Triphasic Levonorgestrel Combination." *Acta Obstetricia et Gynecologica Scandinavica* 66, suppl. 143 (1987): 41–44.

Koutras, D. A. "Disturbances of Menstruation in Thyroid Disease." *Annals of the New York Academy of Sciences* 816 (1997): 280–84.

Krassas, G. E. "Thyroid Disease and Female Reproduction." *Fertility and Sterility* 74, no. 6 (2000): 1063–70.

Lauritano, E., et al. "Association Between Hypothyroidism and Small Intestinal Bacterial Overgrowth." *Journal of Clinical Endocrinology and Metabolism* 92, no. 11 (2007): 4180–84.

Mackawy, A. M. H., et al. "Vitamin D Deficiency and Its Association with Thyroid Disease." *International Journal of Health Sciences* 7, no. 3 (2013): 267–75.

Mansournia, N., et al. "The Association Between Serum 25OHD Levels and Hypothyroid Hashimoto's Thyroiditis." *Journal of Endocrinological Investigation* 37, no. 5 (2014): 473–76.

Olsson, E. M. G., et al. "A Randomised, Double-Blind, Placebo-Controlled, Parallel-Group Study of the Standardised Extract SHR-5 of the Roots of *Rhodiola Rosea* in the Treatment of Subjects with Stress-Related Fatigue." *Planta Medica* 75 (2009): 105–12.

Ongphiphadhanakul, B., et al. "Tumor Necrosis Factor-Alpha Decreases Thyrotropin-Induced 5'-Deiodinase Activity in FRTL-5 Thyroid Cells." *European Journal of Endocrinology* 130, no. 5 (1994): 502–7.

Palmery, M., et al. "Oral Contraceptives and Changes in Nutritional Requirements." *European Review for Medical and Pharmacological Sciences* 17, no. 13 (2013): 1804–13.

Peterson, A. M., et al. "The Anti-inflammatory Effect of Exercise." *Journal of Applied Physiology* 98, no. 4 (2005): 1154–62.

Ramya, M. R., et al. "Menstrual Disorders Associated with Thyroid Dysfunction." *International Journal of Reproduction, Contraception, Obstetrics, and Gynecology* 6, no. 11 (2017): 5113–17. http://www.ijrcog.org/index.php/ijrcog/article/viewFile /3718/2946.

Rettori, V., et al. "Central Action of Interleukin-1 in Altering the Release of TSH, Growth Hormone, and Prolactin in the Male Rat." *Journal of Neuroscience Research* 18, no. 1 (1987): 179–83.

Sapolsky, R. M., et al. "The Neuroendocrinology of Stress and Aging: The Glucocorticoid Cascade Hypothesis." *Endocrine Reviews* 7, no. 3 (1986): 284–301.

Sørensen, C. J., et al. "Combined Oral Contraception and Obesity Are Strong Predictors of Low Grade Inflammation in Healthy Individuals: Results from the Danish Blood Donor Study (DBDS)." *PLoS One* 9, no. 2 (2014): e88196.

Stagnaro-Green, A. "Approach to the Patient with Postpartum Thyroiditis." *Journal of Clinical Endocrinology and Metabolism* 97, no. 2 (2012): 334–42.

Steingold, K. A., "Comparison of Transdermal to Oral Estradiol Administration on Hormonal Hepatic Parameters in Women with Premature Ovarian Failure." *Journal of Clinical Endocrinology and Metabolism* 73, no. 2 (1991): 275–80.

Stojanovich, L., et al. "Stress as a Trigger of Autoimmune Disease." *Autoimmunity Reviews* 7, no. 3 (2008): 209–13.

Surks, M. I., et al. "Drugs and Thyroid Function." *New England Journal of Medicine* 333, no. 25 (1995): 1688–94.

Thorp, V. J. "Effect of Oral Contraceptive Agents on Vitamins and Mineral Requirements." *Journal of the American Dietetic Association* 76, no. 6 (1980): 581–84.

Verlarde-Mayol, C., et al. "Pernicious Anemia and Autoimmune Thyroid Disease in Elderly People." [In Spanish.] *Revista Española de Geriatría y Gerontología* 50, no. 3 (2015): 126–28.

Walter, K. N., et al. "Elevated Thyroid Stimulating Hormone Is Associated with Elevated Cortisol in Healthy Young Men and Women." *Thyroid Research* 5 (2012): 13.

Webb, J. L. "Nutritional Effects of Oral Contraceptive Use: A Review." *Journal of Reproductive Medicine* 25, no. 4 (1980): 150–56.

Westhoff, C., et al. "Using Changes in Binding Globulins to Assess Oral Contraceptive Compliance." *Contraception* 87, no. 2 (2013): 176–81.

Woodruff, S. C., et al. "Mindfulness and Anxiety." In *The Wiley Blackwell Handbook of Mindfulness,* edited by A. Ie, C. T. Ngnoumen, and E. J. Langer, ch. 37. New York: John Wiley, 2014.

Chapter 8: Reverse Metabolic Mayhem

Abdollahi, M., et al. "Obesity: Risk of Venous Thrombosis and the Interaction with Coagulation Factor Levels and Oral Contraceptive Use." *Thrombosis and Haemostasis* 89, no. 3 (2003): 493–98.

Alvarez, J. A., et al. "Role of Vitamin D in Insulin Secretion and Insulin Sensitivity for Glucose Homeostasis." *International Journal of Endocrinology* (2010): 351385.

Bakir, R., et al. "Lipids, Lipoproteins, Arterial Accidents, and Oral Contraceptives." [In French.] *Contraception, Fertilité, Sexualité* 14, no. 1 (1986): 81–87.

Bloemenkamp, K. W., et al. "Risk of Venous Thrombosis with Use of Current Low-Dose Oral Contraceptives Is Not Explained by Diagnostic Suspicion and Referral Bias." *Archives of Internal Medicine* 159, no. 1 (1999): 65–70.

Bultman, S. J. "Emerging Roles of the Microbiome in Cancer." *Carcinogenesis* 35, no. 2 (2014): 249–55.

Bushnell, C. D. "Stroke in Women: Risk and Prevention Throughout the Lifespan." *Neurologic Clinics* 26, no. 4 (2008): 1161-xi.

Cauci, S., et al. "Effects of Third-Generation Oral Contraceptives on High-Sensitivity C-Reactive Protein and Homocysteine in Young Women." *Obstetrics and Gynecology* 111, no. 4 (2008): 857–64.

Centers for Disease Control. "Long-Term Oral Contraceptive Use and the Risk of Breast Cancer. The Centers for Disease Control Cancer and Steroid Hormone Study." *JAMA* 249, no. 12 (1983): 1591–95.

Chasan-Taber, L., et al. "Epidemiology of Oral Contraceptives and Cardiovascular Disease." *Annals of Internal Medicine* 128, no. 6 (1998): 467–77.

Chen, L., et al. "Mechanisms Linking Inflammation to Insulin Resistance." *International Journal of Endocrinology* (2015): 508409.

Cummings, S. R., et al. "Prevention of Breast Cancer in Postmenopausal Women: Approaches to Estimating and Reducing Risk." *Journal of the National Cancer Institute* 101, no. 6 (2009): 384–98.

De Bastos, M., et al. "Combined Oral Contraceptives: Venous Thrombosis." *Cochrane Database of Systematic Reviews* 3, no. 3 (2014): CD010813.

De Bruijn, S. F., et al. "Case-Control Study of Risk of Cerebral Sinus Thrombosis in Oral Contraceptive Users and in [Correction of Who Are] Carriers of Hereditary Prothrombotic Conditions. The Cerebral Venous Sinus Thrombosis Study Group." *British Medical Journal* 316, no. 7131 (1998): 589–92.

Diab, K. M., et al. "Contraception in Diabetic Women: Comparative Metabolic Study of Norplant, Depot Medroxyprogesterone Acetate, Low Dose Oral Contraceptive Pill and CuT380A." *Journal of Obstetrics and Gynaecology Research* 26, no. 1 (2000): 17–26.

Dinger, J. C., et al. "The Safety of a Drospirenone-Containing Oral Contraceptive: Final Results from the European Active Surveillance Study on Oral Contraceptives Based on 142,475 Women—Years of Observation." *Contraception* 75, no. 5 (2007): 344–54.

Duarte, C., et al. "Oral Contraceptives and Systemic Lupus Erythematosus: What Should We Advise to Our Patients?" *Acta Reumatológica Portuguesa* 35, no. 2 (2010): 133–40.

European Medicines Agency. "PhVWP Monthly Report on Safety Concerns, Guidelines, and General Matters." No. 1201 (January 2012). http://www.ema.europa.eu/docs/en_GB /document_library/Report/2012/01/WC500121387.pdf.

Fang, X., et al. "Dose-Response Relationship Between Dietary Magnesium Intake and Cardiovascular Mortality: A Systemic Review and Dose-Based Meta-Regression Analysis of Prospective Studies." *Journal of Trace Elements in Medicine and Biology* 38 (2016): 64–73.

Glintborg, D. "Increased Thrombin Generation in Women with Polycystic Ovary Syndrome: A Pilot Study on the Effect of Metformin and Oral Contraceptives." *Metabolism* 64, no. 10 (2015): 1272–78.

Greenlund, K. J., et al. "Associations of Oral Contraceptive Use with Serum Lipids and Lipoproteins in Young Women: The Bogalusa Heart Study." *Annals of Epidemiology* 7, no. 8 (1997): 561–67.

Haarala, A., et al. "Use of Combined Oral Contraceptives Alters Metabolic Determinants and Genetic Regulation of C-Reactive Protein. The Cardiovascular Risk in Young Finns Study." *Scandinavian Journal of Clinical and Laboratory Investigation* 69, no. 2 (2009): 168–74.

Hankinson, S. E., et al. "A Prospective Study of Oral Contraceptive Use and Risk of Breast Cancer (Nurses' Health Study, United States)." *Cancer Causes and Control* 8, no. 1 (1997): 65–72.

Houston, M. "The Role of Magnesium in Hypertension and Cardiovascular Disease." *Journal of Clinical Hypertension* 13, no. 11 (2011): 843–47.

Kalkhoff, R. K. "Effects of Oral Contraceptive Agents and Sex Steroids on Carbohydrate Metabolism." *Annual Review of Medicine* 23 (1972): 429–38.

Kemmeren, J. M., et al. "Effect of Second- and Third-Generation Oral Contraceptives on the Protein C System in the Absence or Presence of the Factor V Leiden Mutation: A Randomized Trial." *Blood* 103, no. 3 (2004): 927–33.

Kemmeren, J. M., et al. "Effects of Second- and Third-Generation Oral Contraceptives and Their Respective Progestagens on the Coagulation System in the Absence or Presence of the Factor V Leiden Mutation." *Thrombosis and Haemostasis* 87, no. 2 (2002): 199–205.

Kim, Sung-Woo, et al. "Long-Term Effects of Oral Contraceptives on the Prevalence of Diabetes in Post-menopausal Women: 2007–2012 KNHANES." *Endocrine* 53, no. 3 (2016): 816–22.

Kjos, S. L., et al. "Contraception and the Risk of Type 2 Diabetes Mellitus in Latina Women with Prior Gestational Diabetes Mellitus." *JAMA* 280, no. 6 (1998): 533.

Krauss, R. M., et al. "The Metabolic Impact of Oral Contraceptives." *American Journal of Obstetrics and Gynecology* 167, no. 4, pt. 2 (1992): 1177–84.

Kuhl, H. "Hormonal Contraception and Substitution Therapy: The Importance of Progestogen for Cardiovascular Diseases." [In German.] *Geburtshilfe und Frauenheilkunde* 52, no. 11 (1992): 653–62.

Larivée, N., et al. "Drospirenone-Containing Oral Contraceptive Pills and the Risk of Venous Thromboembolism: An Assessment of Risk in First-Time Users and Restarters." *Drug Safety* 40, no. 7 (2017): 583–96.

Lizarelli, P. M., et al. "Both a Combined Oral Contraceptive and Depot Medroxyprogesterone Acetate Impair Endothelial Function in Young Women." *Contraception* 79, no. 1 (2009): 35–40.

Lobo, R. A., et al. "The Importance of Diagnosing the Polycystic Ovary Syndrome." *Annals of Internal Medicine* 132, no. 12 (2000): 989–93.

Lydic, M. L., et al. "Chromium Picolinate Improves Insulin Sensitivity in Obese Subjects with Polycystic Ovary Syndrome." *Fertility and Sterility* 86, no. 1 (2006): 243–46.

Macik, B. G., et al. "Thrombophilia: What's a Practitioner to Do?" *Hematology, American Society of Hematology Education Program* (2001): 322–38. http://asheducationbook .hematologylibrary.org/content/2001/1/322.full.pdf.

Marchbanks, P. A., et al. "Oral Contraceptives and the Risk of Breast Cancer." *New England Journal of Medicine* 346 (2002): 2025–32.

Martinelli, I., et al. "High Risk of Cerebral-Vein Thrombosis in Carriers of a Prothrombin-Gene Mutation and in Users of Oral Contraceptives." *New England Journal of Medicine* 338, no. 25 (1998): 1793–97.

Martinelli, I., et al. "Hyperhomocysteinemia in Cerebral Vein Thrombosis." *Blood* 102, no. 4 (2003): 1363–66.

Martinelli, I., et al. "Interaction Between the G20210A Mutation of the Prothrombin Gene and Oral Contraceptive Use in Deep Vein Thrombosis." *Arteriosclerosis, Thrombosis, and Vascular Biology* 19, no. 3 (1999): 700–3.

Mavropoulos, J. C., et al. "The Effects of a Low-Carbohydrate, Ketogenic Diet on the Polycystic Ovary Syndrome: A Pilot Study." *Nutrition and Metabolism* 2 (2005): 35.

Merki-Feld, G. S. "Effect of Combined Hormonal Contraceptives on the Vascular Endothelium and New Cardiovascular Risk Parameters." [In German.] *Therapeutische Umschau* 66, no. 2 (2009): 89–92.

Middeldorp, S., et al. "Effects on Coagulation of Levonorgestrel- and Desogestrel-Containing Low Dose Oral Contraceptives: A Cross-Over Study." *Thrombosis and Haemostasis* 84, no. 1 (2000): 4–8.

Moghadam, A. M., et al. "Efficacy of Omega-3 Fatty Acid Supplementation on Serum Levels of Tumour Necrosis Factor-Alpha, C-Reactive Protein and Interleukin-2 in Type 2 Diabetes Mellitus Patients." *Singapore Medical Journal* 53, no. 9 (2012): 615–19.

Mohllajee, A. P., et al. "Does Use of Hormonal Contraceptives Among Women with Thrombogenic Mutations Increase Their Risk of Venous Thromboembolism? A Systematic Review." *Contraception* 73, no. 2 (2006): 166–78.

Moreno, V., et al. "Effect of Oral Contraceptives on Risk of Cervical Cancer in Women with Human Papillomavirus Infection: The IARC Multicentric Case-Control Study." *Lancet* 359, no. 9312 (2002): 1085–92.

Murray, E. K. I., et al. "Thromboelastography Identifies Cyclic Haemostatic Variations in Healthy Women Using Oral Contraceptives." *Thrombosis Research* 136, no. 5 (2015): 1022–26.

Nestler, J. E., et al. "Ovulatory and Metabolic Effects of D-chiro-inositol in the Polycystic Ovary Syndrome." *New England Journal of Medicine* 340, no. 17 (1999): 1314–20.

Nightingale, A. L., et al. "The Effects of Age, Body Mass Index, Smoking, and General Health on the Risk of Venous Thromboembolism in Users of Combined Oral Contraceptives." *European Journal of Contraception and Reproductive Health Care* 5, no. 4 (2000): 265–74.

Nordio, M., et al. "The Combined Therapy with Myo-inositol and D-chiro-inositol Reduces the Risk of Metabolic Disease in PCOS Overweight Patients Compared to Myo-inositol Supplementation Alone." *European Review for Medical and Pharmacological Sciences* 16, no. 5 (2012): 575–81.

Oner, G., et al. "Clinical, Endocrine, and Metabolic Effects of Metformin vs. N-Acteylcytseine in Women with Polycystic Ovary Syndrome." *European Journal of Obstetrics, Gynecology, and Reproductive Biology* 159, no. 1 (2011): 127–31.

"Oral-Contraceptive Use and the Risk of Breast Cancer. The Cancer and Steroid Hormone Study of the Centers for Disease Control and the National Institute of Child Health and Human Development." *New England Journal of Medicine* 315, no. 7 (1986): 405–11.

Orio, F., et al. "Metabolic and Cardiovascular Consequences of Polycystic Ovary Syndrome." *Minerva Ginecologica* 60, no. 1 (2008): 39–51.

Pelusi, B., et al. "Type 2 Diabetes and the Polycystic Ovary Syndrome." *Minerva Ginecologica* 56, no. 1 (2004): 41–51.

Pomp, E. R., et al. "Risk of Venous Thrombosis: Obesity and Its Joint Effect with Oral Contraceptive Use and Prothrombotic Mutations." *British Journal of Haematology* 139, no. 2 (2007): 289–96.

Raghuramulu, N., et al. "Vitamin D Improves Oral Glucose Tolerance and Insulin Secretion in Human Diabetes." *Journal of Clinical Biochemistry and Nutrition* 13, no. 1 (1992): 45–51.

Raitakari, M., et al. "Distribution and Determinants of Serum High-Sensitive C-Reactive Protein in a Population of Young Adults: The Cardiovascular Risk in Young Finns Study." *Journal of Internal Medicine* 258, no. 5 (2005): 428–34.

Roach, R. E., et al. "Combined Oral Contraceptives: The Risk of Heart Attack and Stroke in Women Using Birth Control Pills." *Cochrane Database of Systematic Reviews* no. 8 (2015): CD011054.

Roberfroid, M., et al. "Prebiotic Effects: Metabolic and Health Benefits." *British Journal of Nutrition* 104, suppl. 2 (2010): S1–63.

Rocha, S. L., et al. "Heart Attack in a Young Woman—All About Genetics!" *Journal of US-China Medical Science* 14 (2017): 28–30. https://www.davidpublisher.com/Public/uploads/Contribute/58b7979d5aba9.pdf.

Rosing, J., et al. "Low-Dose Oral Contraceptives and Acquired Resistance to Activated Protein C: A Randomised Cross-over Study." *Lancet* 354, no. 9195 (1999): 2036–40.

Rosing, J., et al. "Oral Contraceptives and Venous Thrombosis: Different Sensitivities to Activated Protein C in Women Using Second- and Third-Generation Oral Contraceptives." *British Journal of Haematology* 97, no. 1 (1997): 233–38.

Samsioe, G. "Coagulation and Anticoagulation Effects of Contraceptive Steroids." *American Journal of Obstetrics and Gynecology* 170, no. 5, pt. 2 (1994): 1523–27.

Samsunnahar, Q. S. A., et al. "Assessment of Coagulation Disorder in Women Taking Oral Contraceptives." *Journal of Bangladesh Society of Physiologist* 9, no. 1 (2014): 1–5.

Sasieni, P. "Cervical Cancer Prevention and Hormonal Contraception." *Lancet* 370, no. 9599 (2007): 1591–92.

Seeger, J. D., et al. "Risk of Thromboembolism in Women Taking Ethinylestradiol/Drospirenone and Other Oral Contraceptives." *Obstetrics and Gynecology* 110, no. 3 (2007): 587–93.

Shoelson, S. E., et al. "Inflammation and Insulin Resistance." *Journal of Clinical Investigation* 116, no. 7 (2006): 1793–1801.

Sidney, S., et al. "Venous Thromboembolic Disease in Users of Low-Estrogen Combined Estrogen-Progestin Oral Contraceptives." *Contraception* 70, no. 1 (2004): 3–10.

Singh, B., et al. "Resveratrol Inhibits Estrogen-Induces Breast Carcinogenesis Through Induction of NRF2-Mediated Protective Pathways." *Carcinogenesis* 35, no. 8 (2014): 1872–80.

Skouby, S. O., et al. "Contraception for Women with Diabetes: An Update." *Bailliére's Clinical Obstetrics and Gynaecology* 5, no. 2 (1991): 493–503.

Smith J. S., et al. "Cervical Cancer and Use of Hormonal Contraceptives: A Systematic Review." *Lancet* 361, no. 9364 (2003): 1159–67.

Soares, G. M., et al. "Metabolic and Cardiovascular Impact of Oral Contraceptives in Polycystic Ovary Syndrome." *International Journal of Clinical Practice* 63, no. 1 (2009): 160–69.

Sørensen, C. J., et al. "Combined Oral Contraception and Obesity Are Strong Predictors of Low-Grade Inflammation in Healthy Individuals: Results from the Danish Blood Donor Study (DBDS)." *PLoS One* 9, no. 2 (2014): e88196.

Tayyebi-Khosroshahi, H., et al. "Effect of Treatment with Omega-3 Fatty Acids on C-Reactive Protein and Tumor Necrosis Factor-Alfa in Hemodialysis Patients." *Saudi Journal of Kidney Diseases and Transplantation* 23, no. 3 (2012): 500–6.

Thys-Jacobs, S., et al. "Vitamin D and Calcium Dysregulation in the Polycystic Ovarian Syndrome." *Steroids* 64, no. 6 (1999): 430–35.

Vandenbroucke, J. P., et al. "Increased Risk of Venous Thrombosis in Oral-Contraceptive Users Who Are Carriers of Factor V Leiden Mutation." *Lancet* 344, no. 8935 (1994): 1453–57.

Van Hylckama, V. A., et al. "The Venous Thrombotic Risk of Oral Contraceptives, Effects of Oestrogen Dose and Progestogen Type: Results of the MEGA Case-Control Study." *British Medical Journal* 339 (2009): b2921.

Vasilakis, C., et al. "Risk of Idiopathic Venous Thromboembolism in Users of Progestagens Alone." *Lancet* 354, no. 9190 (1999): 1610–11.

Vasilakis, C., et al. "The Risk of Venous Thromboembolism in Users of Postcoital Contraceptive Pills." *Contraception* 52, no. 2 (1999): 79–83.

Vessey, M., et al. "Oral Contraceptive Use and Cancer. Findings in a Large Cohort Study 1968–2004." *British Journal of Cancer* 95, no. 3 (2006): 385–89.

Vinogradova, Y., et al. "Use of Combined Oral Contraceptives and Risk of Venous Thromboembolism: Nested Case-Control Studies Using the QResearch and CPRD Databases." *British Medical Journal* 350 (2015): h2135.

Wang, Q., et al. "Effects of Hormonal Contraception on Systemic Metabolism: Cross-Sectional and Longitudinal Evidence." *International Journal of Epidemiology* 45, no. 5 (2016): 1445–57.

Westhoff, C. L., et al. "Clotting Factor Changes During the First Cycle of Oral Contraceptive Use." *Contraception* 93, no. 1 (2016): 70–76.

Williams, M. J., et al. "Association Between C-Reactive Protein, Metabolic Cardiovascular Risk Factors, Obesity, and Oral Contraceptive Use in Young Adults." *International Journal of Obesity and Related Metabolic Disorders* 28, no. 8 (2004): 998–1003.

Yan, Y., et al. "Omega-3 Fatty Acids Prevent Inflammation and Metabolic Disorder Through Inhibition of NLRP3 Inflammasome Activation." *Immunity* 38, no. 6 (2013): 1154–63.

Chapter 9: Take Charge of Your Mood Swings, Anxiety, and Depression

Akhondzadeh, S., et al. "Passionflower in the Treatment of Generalized Anxiety: A Pilot Double-Blind Randomized Controlled Trial with Oxazepam." *Journal of Clinical Pharmacy and Therapeutics* 26, no. 5 (2001): 363–67.

Almey, A., et al. "Estrogen Receptors in the Central Nervous System and Their Implication for Dopamine-Dependent Cognition in Females." *Hormones and Behavior* 74 (2015): 125–38.

Anxiety and Depression Association of America. "Facts & Statistics." https://adaa.org/about -adaa/press-room/facts-statistics#.

Arbo, B. D., et al. "Effect of Low Doses of Progesterone in the Expression of the GABA(A) Receptor α4 Subunit and Procaspase-3 in the Hypothalamus of Female Rats." *Endocrine* 46, no. 3 (2014): 561–67.

Arevalo, M. A., et al. "Selective Oestrogen Receptor Modulators Decrease the Inflammatory Response of Glial Cells." *Journal of Neuroendocrinology* 24, no. 1 (2012): 183–90.

Bay-Richter, C., et al. "A Role for Inflammatory Metabolites as Modulators of the Glutamate N-Methyl-D-Aspartate Receptor in Depression and Suicidality." *Brain, Behavior, and Immunity* 43 (2015): 110–17.

Borrow, A. P., et al. "Estrogenic Mediation of Serotonergic and Neurotrophic Systems: Implications for Female Mood Disorders." *Progress in Neuropsychopharmacology and Biological Psychiatry* 54 (2014): 13–25.

Bos, P. A., et al. "Acute Effects of Steroid Hormones and Neuropeptides on Human Social–Emotional Behavior: A Review of Single Administration Studies." *Frontiers in Neuroendocrinology* 33, no. 1 (2012): 17–35.

Bouma, E. M. C., et al. "Adolescents' Cortisol Responses to Awakening and Social Stress; Effects of Gender, Menstrual Phase, and Oral Contraceptives. The TRAILS Study." *Psychoneuroendocrinology* 34, no. 6 (2009): 884–93.

Brinton, R. D., et al. "Progesterone Receptors: Form and Function in Brain." *Frontiers in Neuroendocrinology* 29, no. 2 (2008): 313–39.

Bunevicius, A., et al. "Hypothalamic-Pituitary-Thyroid Axis Function in Women with a Menstrually Related Mood Disorder: Association with Histories of Sexual Abuse." *Psychosomatic Medicine* 74, no. 8 (2012): 810–16.

Carta, M. G., et al. "The Link Between Thyroid Autoimmunity (Antithyroid Peroxidase Autoantibodies) with Anxiety and Mood Disorders in the Community: A Field of Interest or Public Health in the Future." *BMC Psychiatry* 4, no. 1 (2004): 25.

Catuzzi, J. E., et al. "Anxiety Vulnerability in Women: A Two-Hit Hypothesis." *Experimental Neurology* 259 (2014): 75–80.

Daniel, K., et al. "Contraceptive Methods Women Have Ever Used: United States, 1982–2010." *National Health Statistics Reports* 62 (2013): 1–15. https://www.cdc.gov/nchs/data /nhsr/nhsr062.pdf.

Deac, O. M., et al. "Tryptophan Catabolism and Vitamin B-6 Status Are Affected by Gender and Lifestyle Factors in Healthy Young Adults." *Journal of Nutrition* 145, no. 4 (2015): 701–7.

Doufas, A. G., et al. "The Hypothalamic-Pituitary-Thyroid Axis and the Female Reproductive System." *Annals of the New York Academy of Sciences* 900 (2000): 65–76.

Dreon, D. M., et al. "Oral Contraceptive Use and Increased Plasma Concentration of C-reactive Protein." *Life Sciences* 73, no. 10 (2003): 1245–52.

Ervin, K. S., et al. "Estrogen Involvement in Social Behavior in Rodents: Rapid and Long-Term Actions." *Hormones and Behavior* 74 (2015): 53–76.

Fröhlich, M., et al. "Oral Contraceptive Use Is Associated with a Systemic Acute Phase Response." *Fibrinolysis and Proteolysis* 13, no. 6 (1999): 239–44.

Graham, B. M., et al. "Blockade of Estrogen by Hormonal Contraceptives Impairs Fear Extinction in Female Rats and Women." *Biological Psychiatry* 73, no. 4 (2013): 371–78.

Kabat-Zinn, J., et al. "Effectiveness of a Meditation-Based Stress Reduction Program in the Treatment of Anxiety Disorders." *American Journal of Psychiatry* 149, no. 7 (1992): 936–43.

Kessler, R. C. "Epidemiology of Women and Depression." *Journal of Affective Disorders* 74, no. 1 (2003): 5–13.

Klein, S. L., et al. "Sex Differences in Immune Responses." *Nature Reviews Immunology* 16, no. 10 (2016): 626–38.

Kovats, S. "Estrogen Receptors Regulate Innate Immune Cells and Signaling Pathways." *Cellular Immunology* 294, no. 2 (2015): 63–69.

Lakhan, S. E., et al. "Nutritional and Herbal Supplements for Anxiety and Anxiety-Related Disorders: A Systematic Review." *Nutrition Journal* 9 (2010): 42.

Liu, X., et al. "Modulation of Gut Microbiota-Brain Axis by Probiotics, Prebiotics, and Diet." *Journal of Agricultural and Food Chemistry* 63, no. 36 (2015): 7885–95.

Maeng, L. Y., et al. "Sex Differences in Anxiety Disorders: Interactions Between Fear, Stress, and Gonadal Hormones." *Hormones and Behavior* 76 (2015): 106–17.

Mao, J. J., et al. "Rhodiola Rosea Versus Sertaline for Major Depressive Disorder: A Randomized Placebo-Controlled Trial." *Phytomedicine* 22, no. 3 (2015): 394–99.

Meier, T. B., et al. "Kynurenic Acid Is Reduced in Females and Oral Contraceptive Users: Implications for Depression." *Brain, Behavior, and Immunity* 67 (2018): 59–64.

Miller, A. H., et al. "The Role of Inflammation in Depression: From Evolutionary Imperative to Modern Treatment Target." *Nature Reviews: Immunology* 16, no. 1 (2015): 22–34.

Montoya, E. R., et al. "How Oral Contraceptives Impact Social-Emotional Behavior and Brain Function." *Trends in Cognitive Sciences* 21, no. 2 (2017): 125–36.

Nehlig, A. "Effects of Coffee/Caffeine on Brain Health and Disease: What Should I Tell My Patients?" *Practical Neurology* 16, no. 2 (2016): 89–95.

Palmery, M., et al. "Oral Contraceptives and Changes in Nutritional Requirements." *European Review for Medical and Pharmacological Sciences* 17, no. 13 (2013): 1804–13.

Pletzer, B. A., et al. "50 Years of Hormonal Contraception: Time to Find Out, What It Does to Our Brain." *Frontiers in Neuroscience* 8, no. 8 (2014): 256. https://www.frontiersin.org/articles/10.3389/fnins.2014.00256/full.

Quaranta, S., et al. "Pilot Study of the Efficacy and Safety of a Modified-Release Magnesium 250 mg Tablet (Sincromag) for the Treatment of Premenstrual Syndrome." *Clinical Drug Investigation* 27, no. 1 (2007): 51–58.

Rifai, N., et al. "Population Distributions of C-Reactive Protein in Apparently Healthy Men and Women in the United States: Implication for Clinical Interpretation." *Clinical Chemistry* 49, no. 4 (2003): 666–69.

Scheele, D., et al. "Hormonal Contraceptives Suppress Oxytocin-Induced Brain Reward Responses to the Partner's Face." *Social Cognitive and Affective Neuroscience* 11, no. 5 (2016): 767–74.

Skovlund, C. W., et al. "Association of Hormonal Contraception with Depression." *JAMA Psychiatry* 73, no. 11 (2016): 1154–62.

Skovlund, C. W., et al. "Association of Hormonal Contraception with Suicide Attempts and Suicides." *American Journal of Psychiatry* 175, no. 4 (2018): 336–42.

Smith, C., et al. "A Randomised Comparative Trial of Yoga and Relaxation to Reduce Stress and Anxiety." *Complementary Therapies in Medicine* 15, no. 2 (2007): 77–83.

Sorensen, C. J., et al. "Combined Oral Contraception and Obesity Are Strong Predictors of Low-Grade Inflammation in Healthy Individuals: Results from the Danish Blood Donor Study (DBDS)." *PLoS One* 9, no. 2 (2014): e88196.

Stomati, M., et al. "Contraception as Prevention and Therapy: Sex Steroids and the Brain." *European Journal of Contraception and Reproductive Health Care* 3, no. 1 (1998): 21–28.

Ushiroyama, T., et al. "A Case of Panic Disorder Induced by Oral Contraceptive." *Acta Obstetricia et Gynecologica Scandinavica* 71, no. 1 (1992): 78–80.

Van Wingen, G. A., et al. "Gonadal Hormone Regulation of the Emotion Circuitry in Humans." *Neuroscience* 191 (2011): 38–45.

Worly, B. L., et al. "The Relationship Between Progestin Hormonal Contraception and Depression: A Systematic Review." *Contraception* 97, no. 6 (2018): 478–89.

Yoto, A., et al. "Effects of L-Theanine or Caffeine Intake on Changes in Blood Pressure Under Physical and Psychological Stresses." *Journal of Physiological Anthropology* 31 (2012): 28.

Chapter 10: Boost Your Libido and Fertility

Badawy, A., et al. "N-Acetylcysteine and Clomiphene Citrate for Induction of Ovulation in Polycystic Ovary Syndrome: A Cross-over Trial." *Acta Obstetricia et Gynecologica Scandinavica* 86, no. 2 (2007): 218–22.

Baker, F. C., et al. "Circadian Rhythms, Sleep, and the Menstrual Cycle." *Sleep Medicine* 8, no. 6 (2007): 613–22.

Bentzen, J. G., et al. "Ovarian Reserve Parameters: A Comparison Between Users and Non-users of Hormonal Contraception." *Reproductive BioMedicine Online* 25, no. 6: 612–19.

Blackwell. "Oral Contraceptive Pill May Prevent More than Pregnancy: Could Cause Long-Term Problems with Testosterone." *ScienceDaily*. January 5, 2006. www.sciencedaily.com /releases/2006/01/060104232338.htm.

Bonnyns, M., et al. "Thyroid Gland and Female Sexual Function. I. Relation Outside of Pregnancy." [In French.] *Journal de Gynécologie, Obstétrique, et Biologie de la Reproduction* 11, no. 4 (1982): 457–69.

Cheraghi, E., et al. "N-Acetylcysteine Improves Oocyte and Embryo Quality in Polycystic Ovary Syndrome Patients Undergoing Intracytoplasmic Sperm Injection: An Alternative to Metformin." *Reproduction, Fertility, and Development* 28, no. 6 (2016): 723–31.

Coulam, C. B., et al. "Ultrasonographic Predictors of Implantation After Assisted Reproduction." *Fertility and Sterility* 62, no. 5 (1994): 1004–10.

Cutler, W. B., et al. "Sexual Behavior Frequency and Biphasic Ovulatory Type Menstrual Cycles." *Physiology and Behavior* 34, no. 5 (1985): 805–10.

Cutler, W. B., et al. "Sexual Behavior Frequency and Menstrual Cycle Length in Mature Premenopausal Women." *Psychoneuroendocrinology* 4 (1979): 297–309. https://www .athenainstitute.com/sciencelinks/sexualbehaviorfrequency1979.html.

De Flora, S., et al. "Attenuation of Influenza-like Symptomatology and Improvement of Cell-Mediated Immunity with Long-Term N-Acetylcysteine Treatment." *European Respiratory Journal* 10, no. 7 (1977): 1535–41.

Deligdisch, L. "Hormonal Pathology of the Endometrium." *Modern Pathology* 13, no. 3 (2000): 285–94.

Dewailly, D., et al. "The Physiology and Clinical Utility of Anti-Müllerian Hormone in Women." *Human Reproduction Update* 20, no. 3 (2014): 370–85.

Dording, C. M., et al. "A Double-Blind Placebo-Controlled Trial of Maca Root as Treatment for Antidepressant-Induced Sexual Dysfunction in Women." *Evidence-Based Complementary and Alternative Medicine* 2015 (2015): 949036.

Dorr, N. "Fertility Awareness and the Ovulation Method: Natural Birth Control." *Well-Being* 35 (1978): 18–24.

Evans, J. L., et al. "The Molecular Basis for Oxidative Stress-Induced Insulin Resistance." *Antioxidants and Redox Signaling* 7, nos. 7–8 (2005): 1040–52.

Fernando, S., et al. "Melatonin: Shedding Light on Infertility?—A Review of the Recent Literature." *Journal of Ovarian Research* 7 (2014): 98.

Fouany, M. R., et al. "Is There a Role for DHEA Supplementation in Women with Diminished Ovarian Reserve?" *Journal of Assisted Reproduction and Genetics* 30, no. 9 (2013): 1239–44.

Fulghesu, A. M., et al. "N-Acetylcysteine Treatment Improves Insulin Sensitivity in Women with Polycystic Ovary Syndrome." *Fertility and Sterility* 77, no. 6 (2002): 1128–35.

Gingnell, M., et al. "Oral Contraceptive Use Changes Brain Activity and Mood in Women with Previous Negative Affect on the Pill—A Double-Blinded, Placebo-Controlled Randomized Trial of a Levonorgestrel-Containing Combined Oral Contraceptive." *Psychoneuroendocrinology* 38, no. 7 (2013): 1133–44.

Grow, D. R., et al. "Oral Contraceptives Maintain a Very Thin Endometrium Before Operative Hysteroscopy." *Fertility and Sterility* 85, no. 1 (2006): 204–7.

Harvey, S. M. "Female Sexual Behavior: Fluctuations During the Menstrual Cycle." *Journal of Psychosomatic Research* 31, no. 1 (1987): 101–10.

Hvidman, H. W., et al. "Anti-Müllerian Hormone Levels and Fecundability in Women with a Natural Conception." *European Journal of Obstetrics, Gynecology, and Reproductive Biology* 217 (2017): 44–52.

Johnson, L. N., et al. "Antimüllerian Hormone and Antral Follicle Count Are Lower in Female Cancer Survivors and Healthy Women Taking Hormonal Contraception." *Fertility and Sterility* 102, no. 3 (2014): 774–81.e3.

Kallio, S., et al. "Antimüllerian Hormone Levels Decrease in Women Using Combined Contraception Independently of Administration Route." *Fertility and Sterility* 99, no. 5 (2013): 1305–10.

Kovacs, P., et al. "The Effect of Endometrial Thickness on IVF/ICSI Outcome." *Human Reproduction* 18, no. 11 (2003): 2337–41.

Legros, J. J. "Inhibitory Effect of Oxytocin on Corticotrope Function in Humans: Are Vasopressin and Oxytocin Ying-Yang Neurohormones?" *Psychoneuroendocrinology* 26, no. 7 (2001): 649–55.

Lisofsky, N. "Hormonal Contraceptive Use Is Associated with Neural and Affective Changes in Healthy Young Women." *NeuroImage* 134 (2016): 597–606.

Lorenz, T. K., et al. "Interaction of Menstrual Cycle Phase and Sexual Activity Predicts Mucosal and Systemic Humoral Immunity in Healthy Women." *Physiology and Behavior* 152, pt. A (2015): 92–98.

Lorenz, T. K., et al. "Partnered Sexual Activity Moderates Menstrual Cycle–Related Changes in Inflammation Markers in Healthy Women: An Exploratory Observational Study." *Fertility and Sterility* 107, no. 3 (2017): 763–73.e3.

Lorenz, T. K. A., et al. "Sexual Activity Modulates Shifts in TH1/TH2 Cytokine Profile Across the Menstrual Cycle: An Observational Study." *Fertility and Sterility* 104, no. 6 (2015): 1513–21.e4.

Lorenz, T. K., et al. "Testosterone and Immune-Reproductive Tradeoffs in Healthy Women." *Hormones and Behavior* 88 (2017): 122–130.

Magon, N., et al. "The Orgasmic History of Oxytocin: Love, Lust, and Labor." *Indian Journal of Endocrinology and Metabolism* 15, suppl. 3 (2011): S156–61.

Masha, A., et al. "Prolonged Treatment with N-Acetylcysteine and L-Arginine Restores Gonadal Function in Patients with Polycystic Ovary Syndrome." *Journal of Endocrinological Investigation* 32, no. 11 (2009): 870–72.

Montoya, E. R., et al. "How Oral Contraceptives Impact Social-Emotional Behavior and Brain Function." *Trends in Cognitive Sciences* 21, no. 2 (2017): 125–36.

Nassaralla, C. L., et al. "Characteristics of the Menstrual Cycle After Discontinuation of Oral Contraceptives." *Journal of Women's Health* 20, no. 2 (2011): 169–77.

Ohno, Y., et al. "Endometrial Oestrogen and Progesterone Receptors and Their Relationship to Sonographic Appearance of the Endometrium." *Human Reproduction Update* 4, no. 5 (1998): 560–64.

Oner, G., et al. "Clinical, Endocrine, and Metabolic Effects of Metformin vs. N-Acetylcysteine in Women with Polycystic Ovary Syndrome." *European Journal of Obstetrics and Gynecology and Reproductive Biology* 159, no. 1 (2011): 127–31.

Panjari, M., et al. "DHEA for Postmenopausal Women: A Review of the Evidence." *Maturitas* 66, no. 2 (2010): 172–79.

Panzer, C., et al. "Impact of Oral Contraceptives on Sex Hormone–Binding Globulin and Androgen Levels: A Retrospective Study in Women with Sexual Dysfunction." *Journal of Sexual Medicine* 3, no. 1 (2006): 104–13.

Parry, B. L., et al. "Circadian Rhythms of Prolactin and Thyroid-Stimulating Hormone During the Menstrual Cycle and Early Versus Late Sleep Deprivation in Premenstrual Dysphoric Disorder." *Psychiatry Research* 62, no. 2 (1996): 147–60.

Parry, B. L., et al. "Sleep, Rhythms, and Women's Mood. Part I. Menstrual Cycle, Pregnancy, and Postpartum." *Sleep Medicine Reviews* 10, no. 2 (2006): 129–44.

Petersen, K. B., et al. "Ovarian Reserve Assessment in Users of Oral Contraception Seeking Fertility Advice on Their Reproductive Lifespan." *Human Reproduction* 30, no. 10 (2015): 2364–75.

Resnik, S. S. "Melasma and Other Skin Manifestations or Oral Contraceptives." *Transactions of the New England Obstetrical and Gynecological Society* 21 (1967): 101–7.

Smith, G. D., et al. "Sex and Death: Are They Related? Findings from the Caerphilly Cohort Study." *British Medical Journal* 315, no. 7123 (1997): 1641–44.

Soltan, M. H., et al. "Outcome in Patients with Post-Pill Amenorrhoea." *British Journal of Obstetrics and Gynaecology* 89, no. 9 (1982): 745–48.

Talukdar, N., et al. "Effect of Long-Term Combined Oral Contraceptive Pill Use on Endometrial Thickness." *Obstetrics and Gynecology* 120, no. 2, pt. 1 (2012): 348–54. https://journals.lww.com/greenjournal/Fulltext/2012/08000/Effect_of_Long_Term _Combined_Oral_Contraceptive.23.aspx.

Toffoletto, S., et al. "Emotional and Cognitive Functional Imaging of Estrogen and Progesterone Effects in the Female Human Brain: A Systematic Review." *Psychoneuroendocrinology* 50 (2014): 28–52.

Wallwiener, C. W., et al. "Prevalence of Sexual Dysfunction and Impact of Contraception in Female German Medical Students." *Journal of Sexual Medicine* 7, no. 6 (2010): 2139–48.

Zimmerman, Y., et al. "The Effect of Combined Oral Contraception on Testosterone Levels in Healthy Women: A Systematic Review and Meta-Analysis." *Human Reproduction Update* 20, no. 1 (2014): 76–105.

Chapter 13: Alternative Birth Control Methods

American College of Obstetricians and Gynecologists. "Barrier Methods of Birth Control." Last modified March 2018. Accessed on August 28, 2017. https://www.acog.org/Patients /FAQs/Barrier-Methods-of-Birth-Control-Spermicide-Condom-Sponge-Diaphragm -and-Cervical-Cap.

Arévalo, M., et al. "A Fixed Formula to Define the Fertile Window of the Menstrual Cycle as the Basis of a Simple Method of Natural Family Planning." *Contraception* 60, no. 6 (1999): 357–60.

Arévalo, M., et al. "Application of Simple Fertility Awareness-Based Methods of Family Planning to Breastfeeding Women." *Fertility and Sterility* 80, no. 5 (2003): 1241–48.

Arévalo, M., et al. "Efficacy of a New Method of Family Planning: The Standard Days Method." *Contraception* 65, no. 5 (2002): 333–38.

Cervical Barrier Advancement Society. "Contraceptive Efficacy Rates for Cervical Barriers and Other Barriers." Accessed August 28, 2017. http://www.cervicalbarriers.org /information/efficacyRates.htm.

Curtis, K. M., et al. "U.S. Selected Practice Recommendations for Contraceptive Use." *MMWR Recommendations and Reports* 65, no. 4 (2016): 1–66.

Duane, M., et al. "The Performance of Fertility Awareness-Based Method Apps Marketed to Avoid Pregnancy." *Journal of the American Board of Family Medicine* 29, no. 4 (2016): 508–11.

Farley, T. M. M., et al. "Intrauterine Devices and Pelvic Inflammatory Disease: An International Perspective." *Lancet* 339, no. 8796 (1992): 785–88.

Ferrell, R. J., et al. "The Length of Perimenopausal Menstrual Cycles Increases Later and to a Greater Degree than Previously Reported." *Fertility and Sterility* 86, no. 3 (2006): 619–24.

Fihn, S. D., et al. "Association Between Use of Spermicide-Coated Condoms and Escherichia Coli Urinary Tract Infection in Young Women." *American Journal of Epidemiology* 144, no. 5 (1996): 512–20.

Frank-Herrmann, P., et al. "The Effectiveness of a Fertility Awareness Based Method to Avoid Pregnancy in Relation to a Couple's Sexual Behaviour During the Fertile Time: A Prospective Longitudinal Study." *Human Reproduction* 22, no. 5 (2007): 1310–19.

Gallo, M. F., et al. "Cervical Cap Versus Diaphragm for Contraception." *Cochrane Database of Systematic Reviews* 2002, no. 4 (2002): CD003551.

Hatcher, R. A. *Contraceptive Technology.* 20th ed. Decatur, IL: Bridging the Gap Communications, 2011.

Mauck, C., et al. "A Comparative Study of the Safety and Efficacy of FemCap, A New Vaginal Barrier Contraceptive, and the Ortho All-Flex Diaphragm." *Contraception* 60, no. 2 (1999): 71–80.

Sinai, I., et al. "The TwoDay Algorithm: A New Algorithm to Identify the Fertile Time of the Menstrual Cycle." *Contraception* 60, no. 2 (1999): 65–70.

Teal, S. B., et al. "Insertion Characteristics of Intrauterine Devices in Adolescents and Young Women: Success, Ancillary Measures, and Complications." *American Journal of Obstetrics and Gynecology* 213, no. 4 (2015): 515.e1–5.

Trussell, J. "Contraceptive Failure in the United States." *Contraception* 83, no. 5 (2011): 397–404.

Trussell, J., et al. "The Creeping Pearl: Why Has the Rate of Contraceptive Failure Increased in Clinical Trials of Combined Hormonal Contraceptive Pills?" *Contraception* 88, no. 5 (2013): 604–10.

Trussell, J., et al. "Further Analysis of Contraceptive Failure of the Ovulation Method." *American Journal of Obstetrics and Gynecology* 165, no. 6, pt. 2 (1991): 2054–59.

Weschler, T. *Taking Charge of Your Fertility: The Definitive Guide to Natural Birth Control, Pregnancy Achievement, and Reproductive Health, 20th Anniversary Edition.* New York: William Morrow, 2015.

World Health Organization. "A Prospective Multicentre Trial of the Ovulation Method of Natural Family Planning. I. The Teaching Phase." *Fertility and Sterility* 36, no. 2 (1981): 152–58.

ACKNOWLEDGMENTS

Confession. I waited until I was ovulating to write this. I love everyone and the world so hard around ovulation (and I bet you do too) that I wanted to be in this energetic flow as I pay homage to the tribe that made this book possible. Oh what, you thought I could do this alone? I'd like to thank the following people for their incredible support and efforts to bring this book to you:

Bryce Hamrick, my brilliant and sexy AF husband, for loving me fiercely, holding space for me to write endlessly, and for being my ride-or-die guy for life. And we can both be grateful I quit the pill and finally noticed you after ten years of friendship. Ridic.

Bensen (aka the Bringer of Light), thank you for always believing in your mama and for teaching me more about the world than anyone ever has. Your insatiable hunger to understand everything about the human body (including why your mama bleeds) lights my heart up in ways no words can express. I cannot wait to see what the future holds for my period-positive, body-literate boy growing up in the age of the feminine. Watch out, world, because this boy will be unstoppable.

To my amazing sister and brother, Janeen Bidegain and Joseph Owens, we've lived through things that others can't even comprehend. I love you both. Thank you for always being you and loving me.

Kathy and Warren Hamrick, my wonderful in-laws, thank you for your support, for loving me like your own daughter, and for your understanding in the moments I was locked away and not able to play.

Bethany McKenna, seriously, how would my world work without you? You are an amazing chef, inspiration, and hope for all that is to come. You love me, my husband, and my son like family, and we love you right back.

Steph Gaudreau. Dude! You were there in the moments when I thought I just might break and, boy, did we have a good laugh! Thank you for your friendship and relentless commitment to speaking your truth.

Izabella and Michael Wentz for being change makers and believers and for always supporting me in my message. You are incredible friends and an inspiration to the world.

JJ Virgin for never letting me play it small, not even for a minute. You are a true friend. You tell me what I don't want to hear when I need to hear it most.

Thank you, Mallory Leone, Andrea Nakayama, Mariza Snyder, Fiona McCulloch, Amber Spears, Carrie Jones, Bree Argetsinger, Lara Briden, Lara Adler, Magdalena Wszelaki, Alex Carrasco, Joan Rosenberg, Kelly Brogan, Tyna Moore, Sara DeFrancesco, Amy Bader, Maya Shetreat, Katie Wells, Summer Bock, Gabrielle Young, Nicole Beurkens, Natasha Chérie, Trevor Cates, Teri Cochrane, Kellyann Petrucci, Diane Sanfilippo, Christine Schaffner, Robyn Openshaw, Jessica Drummond, Jen Iserloh, Doni Wilson, Christine Faler, Laura Schoenfeld, Nicole Jardim, and Pamela Langenderfer, for being my lady tribe and changing women's medicine for the better.

Tami Lynn Kent, Liliana Barzola, Erika Fayina Marie Reiner, and Kimberly Windstar for teaching me the secrets of taking women's medicine to the next level and the power of the feminine energy, and for helping me attune my energy to create all that I have in this world.

Amy, Ian, Liv . . . I love you guys. Always.

BZ Smith for seeing the light in this little girl and always encouraging me to shine brighter.

Yvonne Macias for teaching her kids to use all five senses to recall the happiest of memories. I prescribe this all the time to patients. Mama wisdom at its finest. Rich Macias for showing me what compassion and love look like in a medical practice.

Carli Webb and Erica Favela for continuously supporting my clinical work and helping my patients achieve better outcomes. Erica, thank you for developing the delicious recipes in this book and for my Rubus Health clinics.

My fellow rebels disrupting medicine in a big way: Dave Asprey, Sayer Ji, Vincent Pedre, Steve Wright, Jordan Reasoner, Joe Rignola, Jason Prall, Michael Roesslein, Kirk Gair, Marc Ryan, Anthony Youn, Jerry Bailey, Karl Krummenacher, Tim Organ, Alex Dunks, Alan Christianson, Shawn Tassone, Dallas Hartwig, Rangan Chatterjee, Rupy Aujla, Kevin Gianni, Pedram Shoji, Dickson Thom, Tom O'Bryan, and Titus Chiu.

My incredible publishing team, Celeste Fine, Jaidree Braddix, and John Maas, for taking a leap of faith and supporting me endlessly through this entire process. My HarperOne editors, Gideon Weil and Sydney Rogers, for always having my back, fighting for my voice, and believing in the message of this book.

Lara Asher, you are nothing short of a MF rock star! You made sure my grammar was in check and those periods were in place and added the little comments of inspiration that were everything I needed to persevere! Girl, you had my back from day one and can I just say, thank gawd Dr. Joe Tatta intro'd us. Thank you, Joe. You're a big reason why this book is here.

Thank you to my patients and online community for trusting in me to develop these protocols and for choosing to heal yourself. I appreciate your support of the mission and being a part of the change in women's medicine. To my reader, you're greater than anything you've ever imagined and I cannot express my gratitude for your support and trust in me. Together we will change women's medicine for the better. I believe in you.

I apologize if I missed anyone who has been a part of my growth, inspiration, and movement. I tried to squeeze everyone in! Know you are loved and appreciated.

INDEX

ABOUT THE AUTHOR

DR. JOLENE BRIGHTEN, NMD, is a functional medicine naturopathic doctor and the founder of Rubus Health, a women's medicine clinic that specializes in women's hormones. She is recognized as an international leading expert in post–birth control syndrome and a pioneer in her exploration of the far-reaching impact and long-term side effects associated with hormonal contraceptives. After many years of clinical practice, she has developed a unique program to support women in preventing and treating PBCS, as well as lowering the risks that the pill has created. Dr. Brighten helps women who are looking to go beyond the pill and balance their hormones, and those who want to stay on the pill without long-term consequences. She has made it her mission to educate women about what that daily dose of hormones could do to their bodies and how to protect themselves while on the pill and coming off the pill.

Dr. Brighten is a bestselling author, speaker, and regular contributor to several online publications, including *MindBodyGreen*, and has been featured in the *New York Post, Forbes, Goop,* and *Fitness.* She is the author of *Healing Your Body Naturally After Childbirth.* She is also a medical adviser for one of the first data-driven apps to offer women personalized birth control recommendations. Dr. Brighten currently resides in Portland, Oregon, with her family.

Visit her online at:

www.drbrighten.com

https://www.facebook.com/drbrighten/

https://www.instagram.com/drjolenebrighten/

https://twitter.com/drbrighten